SCENARIO: BATTLEZONE 2010

Events now gather increased momentum, moving swiftly on multiple levels. At the port city of Shabaz, F-117B *Nighthawks* destroy major communication nodes with precision laser-guided bomb strikes.

From the sea, the Trident submarine *Helena* launches an advanced PAC-5 Tomahawk cruise missile. Other Tomahawks are launched from the *Roosevelt* and from the Aegis cruiser *Benfold*, stationed nearby.

As the high-flying SR-71 spyplane's real-time synthetic aperture radar imagery of the command center is relayed to the *Roosevelt*'s combat information center, analysts study the feed on a wide-screen, high resolution computer monitor.

A structure in the foreground immediately captures their attention. Enhancing the digitized imagery, the analysts note the presence of millimeter wave dish antennas. Although the enemy has tried to conceal the antennas, it is clear that this structure houses a backup tactical communications node.

Two of the carrier's already-launched Tomahawk missiles are re-routed in mid-flight and remotely piloted to their target destinations . . .

Tomorrow's Soldier

The Warriors, Weapons,
and Tactics That Will Win
America's Wars in the
Twenty-first Century

DAVID ALEXANDER

AVON BOOKS ◆ NEW YORK

AVON BOOKS, INC.
1350 Avenue of the Americas
New York, New York 10019

Copyright © 1999 by David Alexander
Published by arrangement with the author
Library of Congress Catalog Card Number: 99-94779
ISBN: 0-380-79502-7
www.avonbooks.com

First Avon Books Printing: August 1999

AVON TRADEMARK REG. U.S. PAT. OFF. AND IN OTHER COUNTRIES, MARCA
REGISTRADA, HECHO EN U.S.A.

Printed in the U.S.A.

WCD   10 9 8 7 6 5 4 3 2 1

To win one hundred victories in one hundred battles is not the acme of skill. To subdue the enemy without fighting: this is the acme of skill.

SUN TZU

• • • • • • • • • •

We are now entering a world of imponderables, and at every stage occasions for self-questioning arise. It is a mistake to look too far ahead. Only one line in the chain of destiny can be handled at a time.

WINSTON CHURCHILL

Contents

Preface

THIS IS A book about tomorrow's warfare and the ways in which tomorrow's soldier will interact with the military weapons systems of the near future. It is a book intended alike for readers with only marginal knowledge of military affairs and readers whose knowledge is more advanced.

The book presumes that most of its readers will have a certain basic familiarity with military terms and procedures and also with fundamentals of geopolitics and recent military history. For those who do not, care has been taken to narrate events, explain terms, and decode acronyms that might be arcane or otherwise hard to fathom without some clarification. For those who still have questions, an appendix of acronyms and a selected glossary are provided at the end of the book.

A key point to bear in mind when reading the book is that its subject is ''tomorrow's'' war, a war in line with the best projections of military forecasters that will probably take the form of a major regional conflict as early as the year 2010 and as late as 2020.

Weapons systems, combat tactics, and other aspects of future warfare have all been linked to this specific time frame, i.e., the book tries to deal principally with those weapons likely to be fielded and actually operational around the year 2010.

However, in some cases, gray areas do exist. Some military technologies are close to maturity while others are not. Certain other military technologies are officially nonexistent or have been outwardly canceled, while secretly and clandestinely they live on as "black" programs. In these instances I have tried to use a best-guess approach toward describing the roles such systems may play in tomorrow's war.

As to conventions, I have used the pronoun "he" when referring to tomorrow's soldier. This is purely a term of convenience, and does not disparage the role women play today and will continue to play tomorrow on the front lines of warfare, as they have most recently done in Operations Desert Fox and Allied Force. The allies of America will usually be referred to as "Coalition" countries, a habit of the Gulf War, which will also be called "Desert Storm" and "war in the Gulf" interchangeably.

As this book goes to press, Operation Desert Fox, a campaign of air strikes directed against Iraq, has recently concluded, while Allied Force action over Kosovo has begun. During the four-day mission in Iraq, United States and British combat aircraft flew an estimated six hundred fifty sorties, while over four hundred twenty-five cruise missiles (more than used in the entire Gulf War) were launched from ships and planes against between eighty-nine and one hundred strategic targets. These targets included suspected chemical, biological or nuclear weapons manufacturing plants, Republican Guard units, military command and control centers and Saddam's lavish presidential palaces. The campaign's stated objectives were to degrade Iraq's ability to manufacture and deploy weapons of mass destruction, to dilute Saddam's conventional military capability, to punish Saddam for denying access to United Nations weapon inspection teams, and to weaken the Iraqi dictator's hold on power—perhaps even to spark a popular uprising that might bring him down.

So far, however, Saddam's mastery over his people remains intact, and while he has lost some of his palaces, he has others, with still more under construction. Of even greater importance, Saddam continues to play his by now familiar hide-and-seek games with weapons inspectors as he pursues his aim of turning Iraq into a regional nuclear power. Meanwhile, Iraqi fighter planes challenging the Coalition no-fly zones that cover some sixty percent of Iraqi airspace demonstrate an ongoing belligerence on the part of his military forces.

By the time you read this it's likely that Coalition forces will have struck at Iraq again and produced the same inconclusive results that past military actions have brought, while action in Kosovo intensifies. More likely still is that the strikes will have taken us one step closer to the inevitable end result: the mobilization of tomorrow's soldier to fight tomorrow's war.

ONE

• • • • • • • • • •

War: Past, Present, and Future

YOU ARE TOMORROW'S soldier. You will play your part in the next major war that America will wage.

This war will be fought sometime within the next two decades, possibly as early as the year 2010. Probably it will be a regional war, one involving an aggressor state challenging the strategic interests of America and her global allies. Also probable is that it will be another Coalition war, similar to Desert Storm. Similar . . . but in many ways very different.

Unlike Desert Storm, this war will likely involve sizable numbers of Coalition casualties. It will be no walk in the desert sun. Among other developments it will see the disappearance of what is now called the "home front." The front lines of tomorrow's war will be everywhere, and everyone will be a combatant.

Western high technology weapons and warfighting tactics have taught the aggressor states of the world an important lesson. But it isn't the lesson that might at first come to mind.

The lesson the aggressor states of the world have learned is that they must invest in new battlefield technologies in order to prevail in the next encounter with Coalition warfighters. They have *not* learned, however, to live in peace with their neighbors, to stop building up banned chemical weapons stockpiles, to stop exporting terrorism, or to stop marching in lockstep to sloganeering madmen who claim to have a direct communication link to God.

While the next war will be regional in theater, it will be global in scope. For the first time in history the American "home front" will become highly vulnerable and will likely become the target of enemy attack.

This attack will materialize in many forms. Terrorists will almost certainly target American cities, launching sudden, unexpected strikes against the dense target populations they offer. But attack may come from other directions. Small, stealthy, and deadly cruise missiles with subkiloton nuclear warheads may be fired from launch platforms such as ships or planes lurking off the coastlines of the continental United States. Virtually impossible to detect or destroy, they could claim a horrifying death toll.

Toxic biological and chemical agents may be introduced into the air and water supplies of New York, Chicago, Los Angeles, and other large American cities. London, Paris, and Rome, among other foreign metropolises, will also be vulnerable. Invisible electronic and information attacks, using computer viruses and digital jamming technologies, may disrupt the fundamentals of life we have all come to depend upon, including electrical power, transportation facilities, and communications of all kinds and at all levels, ranging from telephones to Internet linkages.

In the coming conflict there will be no safe havens, no rear echelons, and no traditional front lines. The battlefield will be located wherever the enemy chooses to carry the fight and wherever our side makes a stand to defend itself. To greater and lesser degrees, we will *all* be combatants and we will *all* be targets. And whether we wear a uniform or not, we will *all* play a role in the deadly game of war that leads to victory or defeat.

We will all, in short, be tomorrow's soldier.

Yet despite these developments, for perhaps the first time in human history, the dominant global power has crafted a military doctrine whose object is not to annihilate the enemy but to knock him out and then attempt to reintegrate him into the international community. Warplanners at the Pentagon call this a "soft kill" strategy. Its objective is to destroy the enemy's ability to wage war by annihilating his warfighting infrastructure, to defang the serpent and render it harmless without having to lop off its head.

Not wanting to produce high casualty rates for national and friendly forces is obviously understandable, but to base a strategy on limiting collateral damage to the enemy's forces is revolutionary. Such a military doctrine runs counter to the avowed aims of every ascendant military power that has ever waged war or strategized about waging war. From the days of Sparta to the present, nothing like it has ever been promulgated. For this reason new warfighting doctrines are said to be an RMA, or a "revolution in military affairs."

The new battlefield that will come into being as a result of this RMA will be a joint digital combined arms combat theater where advanced combat systems will enable war to be waged in three dimensions.

This war will be ninja war, embodying many of the propositions found in the writings of the ancient Chinese sage Sun Tzu, who counseled to be where the enemy believes you are not, to strike where the enemy least expects, and to use military forces with precision accuracy.

It will also be war according to the dictates of the nineteenth-century Prussian strategist Karl von Clausewitz, whose injunction to strike at the enemy's "center of gravity" became a catchphrase of the Gulf War–era newsmedia.

Prelude to Future War: World War Two and the Vietnam War

Historical events tend to run in cycles of action and reaction, thesis and antithesis, yin and yang. The origin of tomorrow's digital and combined arms battlefields lies in the advances in military technology and in the innovations in tactics and strategy driven by the operational realities of World War Two. In the same way, the First World War was a continuation and culmination of innovations in military affairs begun in the nineteenth century, such as those devised by Republican France and the Imperial British during the succession of Napoleonic wars that convulsed much of Europe between 1800 and 1815.

In the same way the twenty-first century will reap the fruits—or the whirlwind, depending on how you look at it—of the technological innovations in warfare that were first developed during the opening decades of the twentieth century.

Stealth, nuclear weapons, radar, jet aircraft, commando warfare, ballistic submarines, terminally guided munitions, and the so-called "flying wing" airframe that matured into the B-2 Stealth bomber, all were either invented or first deployed on the battlefield during the Second World War.

World War Two was the first war in which strategic air power evolved to the stage where it could be used to its full potential. The Blitzkrieg approach to warfare, relying on air superiority, concentrated firepower, and spearheads of fast, maneuverable mechanized armor, also was born then. In time, the Blitzkreig developed into America's AirLand Battle concept.

World War Two was also the first war in which civilians became legitimate targets of planned military attack and in which terror bombing became a "legitimate" tool of warfare. This policy began with the German-launched London bombing blitzes; it was brought back to Germany with the firebombing of Dresden; it culminated in the nuclear destruction of Hiroshima and Nagasaki. It is with us today in terrorist attacks all over the world that use civilians as targets, and will tomorrow lead to the disappearance of the home front entirely.

Perhaps of even greater importance, and with consequences more far-reaching for global history, was the military coming of age in the second world war of a nation not then highly renowned for its military accomplishments. This nation eventually eclipsed the military power potentials of all the other Great Powers, both contemporary and historical.

This nation was the United States which, as another consequence of the second world war, has also emerged as the sole remaining military, economic, and cultural superpower on the planet and will therefore likely play a dominant role in shaping world history for decades, if not centuries, to come.

As was the case with other upheavals in former Great Power colonies, the insurrectionist warfare that led to the war in Vietnam was also in large part an outgrowth of the events of World War Two. In the aftermath of the "Unnecessary War" (as Churchill—who felt Hitler should have been stopped sooner—called it when asked for a name by Franklin Delano Roosevelt), a new and complex set of geopolitical and technological realities quickly materialized.

Among the former were the establishment of two communist superstates, one of which, China, had been a longtime ally of the United States, the other, Russia, was a once and future foe. Both of these countries pursued a policy of supporting revolutionary nationalist movements in smaller Third World states that had been the former colonial satrapies of larger Great Powers prior to the second world war.

Among the new technological realities was the existence of nuclear superweapons in the possession of the United States and the Soviet Union and the escalating threat of global nuclear warfare resulting from their proliferation. The crisis that this posed would not finally ebb until the first years of the century's last decade.

Against this backdrop, America's military warplanners adopted a deliberately cautious strategy of "flexible response" intended to ratchet up the scale of violence by small increments so as not to drive the Soviet Union to take the much-feared nuclear option. If the opposition—be it the Soviets or one of their national surrogates—made a move, the United States countered with a well-orchestrated move of its own. This was flexible response in a nutshell.

Richard Nixon, presiding as U.S. president over much of the Vietnam War—a war that had never been declared by the U.S. Congress—viewed Southeast Asia as a strategic "sideshow" to the Cold War. The aim of U.S. involvement would be a sort of poker player's bluff. The Soviets would learn that the United States could not be cowed, yet America would never act so aggressively as to provoke the Russians into using battlefield nuclear weapons.

In the words of a former photoreconnaissance specialist who processed photographic bomb damage assessments from film shot by cameras in high-flying B-52s: "We were in there to just bomb and napalm the living daylights out of Vietnam and tell the Soviets that if we can't have it, then nobody can."

Whether an oversimplification or not, at the strategic level, the Vietnam War may have in fact been "won" because the United States demonstrated clearly that it could in fact "hang tough" and keep its nerve, a theme repeated to the Soviets throughout the Cold War in overt and covert operations of many kinds. At the theater level, however, the war in Vietnam might have been prosecuted somewhat differently and perhaps more effectively.

The World War Two–style maneuver battles that General William Westmoreland, commander of U.S. combat forces in Vietnam until 1968, was prevented from carrying out seem like more appropriate tactics in light of current revisionist thinking. A concentration of forces at the corps level could have conceivably punched and battered its way into Hanoi and brought the Vietnam War to a quicker conclusion. But early on the die had been cast: the nuclear warfighting strategy of flexible response was misapplied to conventional combat in a regional theater.

Vietnam also inaugurated another feature of modern warfare that will likely stay with us—the "quick-in, quick-out" strategy. When decisive results were not forthcoming, popular national support for foreign military intervention quickly began to wane and demands for pullout predominated. This symptom of the "Vietnam syndrome" has been noted by America's enemies, including Saddam Hussein, in their calculations on how far they dare to go and how much they dare to grab before they face armed retaliation.

The Gulf War and Its Lessons

As the twenty-first century approaches, the geopolitical backdrop to our era has grown to resemble the concluding years of the nineteenth century. Western nations have gone to war in the Mideast to stop a dictator bent on regional conquest (the Mahdist uprising in the 1890s concluding in the battle of Omdurman). Global peace prevails (the long peace following the Allied victory at Waterloo), and commerce as well as military technology have become "globalized." Technological change shapes all spheres of civilian and military life as the Industrial Revolution had done in the previous century.

Also pertinent, though not in a direct chronological sense, is that what was in reality a military campaign with limited results has been elevated to the status of a great and conclusive military victory (Napoleonic France at the battles of Austerlitz, Wagram, and others). I'm referring here to the Gulf War. In the Gulf War, America believed itself to have been "cured" of the "Vietnam syndrome" and to have vindicated its investment in the research and development of high technology weapons since the end of the Vietnam War. The Gulf War was also seen as a vindication of AirLand Battle tactics which called for high-intensity aerial bombing and fast, highly maneuverable assaults on high-value targets.

The truth of such sweeping assertions may be debatable. The war ejected Iraq from Kuwait and preempted an Iraqi takeover of Saudi Arabia and her strategic oil fields that was likely to follow. But it fell shy of occupying Baghdad, ousting Saddam from power and replacing his regime with one friendly to Coalition interests, an end critical to a truly conclusive victory. As a result of these shortcomings, Saddam today continues to plan his next military adventure, and we may have to face him again from a less dominant position.

Further, studies conducted on Gulf War weapon systems found that many of them failed to perform as advertised. A postwar report issued by the U.S. General Accounting Office (GAO) stated that claims of the performance of high-tech weapons were ''overstated, misleading, inconsistent with the best available data, or unverifiable.''*

According to the report, military service branches, arms manufacturers, and their congressional cheerleaders had all spun and sometimes cooked less-than-favorable data outright. For example, F-117A bomb hit tallies were doubled from 40 to 80 percent simply by defining a ''successful strike'' as merely the process of launching a round, as opposed to having actually hit anything.

In the Gulf War against Iraq, America and her Coalition allies faced a second-rate foe prepared to wage a World War One–style fixed position battle. In all cases, except those where elite formations such as the

---

*This and other reports issued by the General Accounting Office can be downloaded from the GAO website at *www.gao.gov*. Other pertinent websites are listed at the back of the book.

picked Republican Guard were engaged, Iraqi combat personnel were poorly trained, poorly equipped, and not strongly motivated to persevere against attack. Even when first-line equipment, such as Soviet-built MiG-29 fighter planes or T-72 tanks were deployed, deficiencies in enemy training overcame the performance capabilities of the equipment used by opposition forces.

Desert Storm was custom-made for U.S. warfighting doctrine. Though putatively the fourth largest in the world, the Iraqi army was organized along rigid Soviet lines, made up largely of ill-trained conscripts deployed at the end of overstretched supply lines, and fielding in the main low-quality Soviet-made weapons.

This was precisely the type of army the Pentagon had envisioned fighting in a World War Three land battle taking place in the Middle East, where Soviet troops and allies such as Iraq were expected to strike for Saudi Arabian oil reserves in the first weeks of fighting. In fact, by a strange coincidence, the U.S. Central Command under General H. Norman Schwarzkopf had completed training exercises in Florida for precisely such a battle only some two weeks before Saddam's armies invaded Kuwait.

The Gulf War's vindication of strategic air power as a substitute for land warfare may also need to be reassessed more soberly. In light of the nature of the enemy we faced, it might not be so surprising that the Desert Wind air campaign proved as effective as it did.

Contrast the devastation wreaked on desert bunker outposts manned by poorly motivated Iraqi troops with other examples of saturation bombing drawn

from military history and the picture suddenly
changes into something quite different.

For instance, during the months-long standoff at
Monte Cassino—a critical obstacle toward Allied
liberation of Rome during World War Two—Allied
forces were pinned down by a determined German
garrison at a monastery occupying the top of a mountain stronghold.

Time after time, the German 1st Parachute Division
entrenched within the monastery was subjected to the
poundings of heavy mortar shells and five-hundred-
pound bombs dropped from B-17 Flying Fortresses,
A-36 fighter-bombers, and B-26 Marauders. Yet
nothing the Allies did could force the enemy's sur-
render. Even the most massive of these bombing attacks, which pulverized the monastery to smoldering
ruins, had little effect on the enemy's will to fight.

It was believed that nothing could survive punishment of such extreme magnitude. But the Germans
not only survived the lethal rain of steel and high
explosive, they used the bomb rubble as effective
cover and were even harder to root out than they had
been before. Only after much Allied blood was shed
in wave after wave of head-on infantry attacks on the
stronghold did Cassino finally fall into friendly hands.

In Vietnam, too, massive high explosive strikes
from B-52 strategic bombers that cruised too high to
be seen or heard (the B-52 is capable of reaching altitudes in excess of 60,000 feet, though it is seldom
flown at that height) often left craters where villages
had stood. Napalm and Agent Orange defoliant turned
once verdant jungle into stretches of surreal lunar
wasteland. But here, too, the Vietnamese communists,
a determined if outgunned and ill-equipped enemy,
dug in and came back to fight again. And there are

other historical examples of the same kind, including recent Serb resistance to NATO air strikes.

The important lesson to be learned from the foregoing is that the spectacular results of tactical bombing in the Gulf War may well turn out to be the exception instead of the rule, a desert mirage leading military planners to a quicksand bog instead of an oasis in America's next regional conflict with a rogue state determined to "go the distance" on tomorrow's battlefield.

A Post—Cold War Arms Glut and Tomorrow's Threats

Today, the military policy of the United States is no longer driven by the fixed-threat deterrent equations that held sway in the bipolar world order existing between 1945 and 1990. At present, U.S. military force structures reflect the needs of a continentally based superpower to project its military power rapidly across the globe against a panoply of well-armed, ideologically determined but far smaller and militarily weaker regional powers.

This dynamic not only encompasses conventional threats but also low intensity conflict (LIC), counterterrorist operations, and peacekeeping operations known as "operations other than war" (OOTW). Operations Desert Storm, Provide Comfort, Just Cause, Restore Hope, Sea Angel, and Typhoon are recent examples of the diversity of military engagements in which the United States has played a leading role.

To meet new strategic challenges, U.S. warplanners have emphasized jointness in training and equipment; smaller, better-trained and better integrated forces;

and the extensive deployment of high-technology weapons and communications systems that integrate computers, robotics, and sensor arrays with conventional explosives, lasers, and nuclear explosives in order to get the edge on larger, dumber, and more poorly equipped enemies.

But the same revolutionary advances in military technology that drive America's arsenal-building are also being disseminated throughout the world. These are now being made available to hostile and renegade regimes at bargain-basement prices. Not only technology but geopolitical factors and economics contribute to this unprecedented state of affairs.

With world peace having broken out, the quantity and scope of arms contracts have been drastically reduced from Cold War levels. It has been projected that as much as 80 percent of top-ranked U.S. defense contractors will disappear by the year 2000, either through collapse or by consolidation with larger companies.

Western Europe's defense firms, many of whom have been doing business since the days of Napoleon and Wellington, are facing similar fiscal realities. And in the former Soviet Union, a government-run defense monolith has fragmented into a handful of large companies. These companies have painfully learned in the last decade that they face extinction unless they switch back to making tanks instead of unsuccessful lines of Winnebagos, refrigerators, microwave ovens, and other consumer goods they tried and failed to market in the 1980s.

On a worldwide basis, defense contractors are faced with a surplus of war materiél. The arms market is saturated with high-quality, low-cost weapons systems. Even first-line systems, such as MiG-29 aircraft, diesel-electric submarines, AWACS aircraft, military surveillance satellites, theater ballistic missile defense and anti-ballistic missile defense systems, are for sale. Stealth and counterstealth technologies, too, are all becoming available to purchasers in quantities never before seen.

The remains of the Soviet Union's military-industrial complex will sell virtually anything to anybody with the hard currency to buy it. The French have historically demonstrated a willingness to sell weapons, and so have the Belgians, at least in the case of small arms, a speciality of theirs. Today both are even more aggressively seeking clients. They are pushing for an end to the United Nations arms embargo against Iraq in order to rearm Saddam, once one of their best clients.

Even more politically motivated weapons dealers, such as the United States, Great Britain, and Israel, to name a few, have relaxed their once stringent standards in the face of economic pressures and emerging geopolitical realities.

All of the above-named suppliers of new first-line weapons systems also provide upgrade programs for legacy combat systems such as F-4 Phantoms and MiG-21s. Hybridization and customization of systems for export to transnational purchasers are also recent trends driving big-ticket arms deals.

In hybridization, elements of a weapon system from one manufacturer—or national entity are combined with elements from another, such as in a recent deal between the French Thompson-CSF and Russian firms to merge French data-processing technology (where Russia is weak) with high-power radar, a Russian strength. Elsewhere, hybrid technology allows a customized MiG fighter plane to fire U.S., Russian, and Western European missiles in the course of a single combat sortie.

Strict export controls once limited the enhancement or upgrading of military systems for export on a case-by-case basis, such as in limiting the kind of missiles available on U.S. F-16 fighter planes to Middle Eastern clients or the then-controversial sale of AWACS aircraft to the Saudis early in the Reagan presidency. This is no longer the way things work in the international arms business.

Today, major-league purchasers such as Saudi Arabia balk at systems that do not include first-string weapons, and a recent sale of fighters to the Saudis would not have gone through had not the United States backed down on its original pledge not to sell sophisticated advanced medium range air-to-air missiles (AMRAAM) along with the fighter planes.

The upshot of the new approach to global arms sales of major combat systems is that increasingly more sophisticated weaponry is flowing into the military arsenals of belligerent regimes such as Iraq, Communist China (PRC), North Korea, Zaire, and Iran, to name but a few eager purchasers. It's also finding its way to "Third Force" entities—nonstate actors such as terrorist groups, narcocriminal and mafia organizations (e.g., the Colombian drug cartels and Italian *Stidda*)—and "neo-"organizations of all kinds, which often have ties to multinational corporations and even national governments.

Combined with these forces is a revolution in global mass communications that has "wised up" formerly backward-thinking adversaries who would not know stealth from a camel's hump.

The success of the military approaches used in Desert Storm was also a wake-up call to the rogue nations of the world who have learned the painful lesson that he who does not move forward will inevitably fall behind. These nations have also engaged in a thorough restructuring of military doctrine and force integration since the Gulf conflict and will be far better prepared to challenge the allied powers, either singly or in coalition warfare, in future confrontations.

Our potential enemies have realized a key point from the Gulf War: that they do not need to defeat the U.S. military per se in order to gain an important objective. All they need is to seize a strategically vital piece of territory and thereby force the U.S. and her allies into either fighting a major military engagement or backing down and negotiating. In the latter case, they could still win. Iraq's refusal to give American weapon inspectors access to suspected arms depots in November of 1997 and December of 1998 is a recent strategy based along these lines. There will be more such power plays in the future.

Tomorrow, the proliferation of accurate long-range and beyond visual range (BVR) missiles such as Russian SA-10 and U.S. AMRAAM clones from third-party sources means that U.S. military deployment to distant trouble spots may be more difficult than ever before. As long as the enemy thinks small and keeps the war in his own backyard, there is the distinct possibility of his prevailing in the fight.

The net result of these new and emerging power dynamics is that tomorrow's soldier will in all probability face a foe better armed, better prepared, and less willing to lay down his arms than any regional enemy he's faced beforehand. This enemy of tomorrow will be better able to project his own military power in new and dangerous ways, for instance by using ballistic and stealth technology to attack long-range targets such as U.S. cities with nuclear or high-yield conventional weapons.

A recent war game engineered by DARPA, the U.S. Defense Advanced Research Projects Agency, called Winter War Game 1997, can serve as a cautionary illustration. In this war scenario staged by a strategy assessment program called Army After Next and set in the year 2020, the United States was pitted against several unnamed adversaries in a "two half-wars" scenario, two half or regional wars being the level of aggression that the U.S. currently believes it needs to maintain a constant military readiness to fight.

In the first hours of this computerized warfare scenario, America's mock enemies went straight for the jugular, knocking out the United States' network of military surveillance satellites that serve as the eyes and ears of its armed forces. The enemy used smart, cheap, and sophisticated ground-to-space and air-to-space weapons to kill the satellites, in effect rendering U.S. forces deaf, dazed, and blind.

Pentagon warplanners have characterized the results of this war game as a wake-up call, as well they should. The extreme vulnerability to attack of U.S. space resources has never been seriously questioned, nor have land, sea, and air commanders fully appreciated the extent to which command, control, and communications (C3) are vulnerable to a preemptive strike.

In another study of future war conducted by the British Army, called BA 2000, two basic forms or "views" of a major conflict on which to base doctrine and force structures in the next century were postulated by the British Ministry of Defense (MOD).

View One was termed a "heavy metal" confliction. These were high-intensity operations on a high-technology battlefield similar to, but more advanced than, the Desert Storm campaign.

In View Two, military operations such as those conducted in Bosnia, Chechenya, Afghanistan, Northern Ireland, and Somalia were encompassed. These were unconventional conflicts involving civil disorder and separatist fighting, showing elements of guerrilla warfare combined with heavy urban combat.

The British MOD study concluded that the first type of conflict was the kind most likely to be encountered during the next war, but that the second type was also a salient possibility that had to be considered to maintain a credible level of military preparedness.

U.S. warplanners are relying on an overwhelming technological and tactical superiority to posture America's military forces at a strategic advantage to her adversaries, no matter what mask the god of war decides to don in his next Apocalyptic appearance. Therefore for every major weapon system that trickles down to the minor leagues, America's defense doctrine calls for a new and radical system to be put in place to challenge it. For every future warfighting contingency, battlefield systems are being developed to support flexible doctrinal concepts.

All three arms of the U.S. military are engaged in major efforts to develop these means of waging future war. The Army's program is called Force XXI; the Navy is restructuring along the lines of a plan it terms Copernicus; and the Air Force is pursuing its own vision of future combat it calls Global Engagement.

The United States and her Coalition partners believe that they now possess the technological means to wage tomorrow's war and prevail against any foe likely to challenge them. They also have an awareness

of the types of combat scenarios that military forces
are likely to confront in this future war. But the un-
answered questions are perhaps the biggest and most
important to us all—where and when and why will
this next war happen?

In the next chapter, we will examine some of the
best guesstimates about the flashpoints that may ignite
the tinderbox of geopolitics into the bonfire of Mars,
and thereby bring about tomorrow's war.

TWO

●●●●●●●●●

Flashpoints of Future War

THAT TOMORROW'S SOLDIER will see combat is a certainty. Sometime within the next ten to twenty years—by the year 2010, if some Pentagon predictions are accurate—the strategic interests of America and her allies will be threatened and the grim reaper of war will again mount his pale horse.

Tomorrow's soldier—the universal soldier—will be called on to serve his country as he has done for centuries, and his blood will again stain the sand of the desert or the soil of the jungle or the grass of the woodlands or the bomb-shattered pavements of burning city streets. No matter how high-tech warfare becomes, this much is a certainty.

But where will this future soldier see action? And what will be the incandescent tip of conflict that ignites the next military flashpoint?

What are the likeliest possibilities for regional and Big Power confrontation that might erupt into full-blown hostilities on land, on sea, or in the air?

We can identify some crises in the making, and the interlocking networks of power dynamics that flow like hot magma beneath the sleeping volcanos of the New World Order. The following is a list of the best candidates for where tomorrow's battlefield will likely materialize.

The Beijing-Moscow Axis

China is an emerging military and economic power that some observers believe will play a major role in shaping the course of the twenty-first century. China's economy is still largely agrarian and its brand of old-style party communism shows a broad disdain for human rights, as displayed in the bloody put-down of the uprising for democratic reform in Tiananmen Square on June 3 and 4, 1989 by troops of the ironically-named People's Liberation Army or PLA.

More recently, on July 1, 1997, China regained control of the formerly independent democratic island state of Hong Kong and immediately began to clamp down restrictions on its population.

China is a nation with military aspirations, and Hong Kong is not the only piece of real estate that the Chinese desire. The Chinese have also made territorial claims on the Kuril Islands and other strategic islands in the South China Sea, a body of water lying between the Philippines and Japan. Were China to gain possession of any or all of these islands, she would pose a direct threat to her longtime enemy and current strong U.S. ally in the Far East—Japan.

China also has made territorial claims against Taiwan, which it considers a breakaway province, and two other nearby islands, Kinmen and Matsu. In short, the Chinese aspire to the domination of East Asia and have repeatedly stated that they consider the United States their main opposition to this strategic aim.

"By early in the next century, about the time when

the size of the Chinese economy is projected to surpass our own," said House National Security Committee chairman Floyd Spence in congressional testimony last year, "China will have developed large and capable military forces."

Most recently China has engaged in fence-mending with another of its ancient enemies—Russia. One significant result of this new relationship between Beijing and Moscow is that Russian arms and state-of-the-art military technology has been flowing into China at a record pace.

In 1996, roughly 30 percent of the $3.5 billion that Russia collected on foreign arms sales came from purchases made by China. Moscow plans to increase arms sales to China by $2 billion annually and to increase the level of high technology incorporated into the exported weapons systems.

China has made it clear that she will shop elsewhere unless the former Soviets transfer first-line war technology. The Russians, convinced that their defense sector is the only viable economic sector and the one which will drive Russian recovery, have been unable to refuse to do the deal.

The new Chinese-Russian relationship is not based simply on economics. There are cultural and political links being forged as well. Both countries, with huge populations and vast land masses, envision themselves as major players on the global scene. Russia, a former superpower, can look back nostalgically to a time when she fielded the world's largest army and navy.

Each nation has developed an inferiority complex over what it perceives as the predominant role played by the United States in the so-called New World Order. Many political leaders inside the former Soviet Union (FSU) also are embittered by the expansion of

NATO to include former Warsaw Pact countries.

To many observers, it appears Russia may be in the process of reorienting its interests toward the East instead of the West, where its cultural and political aspirations were formerly focused. Although Russia and China pledged to work together toward a "multipolar world" in a recent state visit to Moscow by Chinese president Jiang Zemin, China's extensive armament by Russia bespeaks a far different and more sinister direction.

Some estimates see a minimum of one hundred joint Sino-Soviet military research and development goals currently being conducted toward such projects as new generation solid-fuel theater ballistic missiles, antiballistic missile defense systems, and new and powerful radars reputedly capable of detecting stealth aircraft.*

Such a buildup of Chinese military forces is not necessary merely for defensive purposes. But it is strategically essential if a country is interested in waging war in pursuit of expansionist ends.

China isn't engaged only in a strategic partnership with the Russian Commonwealth. Beijing has shown a willingness to pay for military technology and arms from any source that has what it needs, including clandestine and gray market channels. Israel has recently joined the list of those supplying China with sophisticated missile technology.

The Chinese are also developing some home-rolled weapons programs. One of the most worrisome, from the standpoint of the West, may be its effort to

---

*See the study "Nuclear Weapons in a Transformed World," Center for Strategic and International Studies, Washington, DC, 1998.

develop multiple independently targetable reentry vehicle (MIRV) technology. This research and development (R&D) has been ongoing since the early 1980s.

MIRVing warheads increases the killing power of a ballistic missile many times over. Instead of a single warhead targeted at a single objective, MIRVed warheads can strike multiple targets after disengaging from the final stage or missile "bus" during the exoatmospheric stage, at the end of the so-called "boost phase" of a ballistic missile's trajectory.

MIRV technology has best application to nuclear weapons delivery systems, and MIRVing nukes was one of the prime forces driving the U.S.-Soviet arms race of the Cold War.

MIRV technology is a destabilizing factor because it fosters a "use 'em or lose 'em" approach to defense. An arsenal of MIRVed missiles invites a first strike because of the tempting capability to destroy many enemy missiles in one shot. The flip side of the coin—a perception of vulnerability to the other side's MIRVed nukes—also drives a strike-first mentality.

As for the former Soviets, the Russians too may have expansionist aims, as reflected in a resurgence of communist sentiment and a nostalgia for the former power and glory of a once-great imperialist land power, fueled by the anxieties brought on by fragmentation and Balkanization of the once-mighty U.S.S.R. Like a collapsing star, the "new" Russia may one day reach a point where it can no longer shrink and begin moving in the opposite direction— with explosive force.

Under Mikhail Gorbachev and Boris Yeltsin, Russia's outer satellites and its Baltic buffer states were allowed to splinter away with impunity early on in the Soviet breakup. But when the Chechens—an ethnic minority group with longstanding grievances against the Russians—decided to stage their own secession from the Russian Commonwealth, the full might of the Russian army was brought to bear against the uprising. The result was a bitter and bloody counterinsurgency war that in the end followed a scorched-earth policy not seen since the brutal closing years of the Second World War.

It's almost a given that China's ascendency in the New World Order, coupled with her avowed territorial hunger and her increasing belligerence, will create new problems for the United States in the coming century.

As the twentieth century closes and the twenty-first begins, the United States finds herself the world's greatest maritime power, plying both Atlantic and Pacific Oceans from her continental base. Unhindered access to those great strategic sea lanes is critical to America's continuing economic and military supremacy. If China were to control the Kuril Islands or other strategic island groups in the Pacific, U.S. power would be gravely threatened.

Similarly, Russian efforts to retake its Baltic buffer states, or any of its former East Bloc client states, would threaten other European countries—especially the reunified Germany—all of whom are strong American allies.

Smoke Clouds in the Middle East

Iran is among those countries that have invested heavily in rearmament and military research and development programs since the end of the Cold War.

Despite having lost the Gulf War and having had United Nations sanctions placed against it, Iraq has also been rearming.

Libya, too, despite the withdrawal of its leader, Mu'ammar al-Gadhafi, from the world scene, and an interest in softening its world image by encouraging high-end tourism (of all things), has been engaging in steady rearmament efforts. And in the neighboring Sudan, Islamic fundamentalists with strong anti-Western agendas have also risen to power in the last ten years.

All of these nations have used both open and clandestine or gray market channels for weapons and technology as sources for their new arsenal of high technology weapons systems. Despite U.N. arms embargoes, technology continually leaks through—up to ten billion dollars' worth in good years, by some estimates.

But while Iraq tries to rearm and upgrade its Scud missile forces despite U.N. weapons inspections and embargoes, and while Syria (which contributed an armored division to the Gulf War Coalition) continues to export terrorism, no single nation in the Mideast poses as much of a threat to global stability as Iran.

Iran's geostrategic position places it at one of the critical crossroads between East and West. Directly across the Persian Gulf from Saudia Arabia and Oman, it borders Turkey and the southwestern portion of the Russian Confederation to the north, Afganistan and Pakistan to the southeast, and Iraq and Syria to the west. Its leaders have expressed designs on the breakaway Azerbaijani republics with the goal of creating an Islamic Iranian superstate that would challenge Russia for dominance in Western Asia.

Iran's dreams of expansionism pose the serious

threat of destabilization in the Mideast, affecting not only Israel but Egypt, Jordan, Saudi Arabia, and Kuwait, all U.S. allies in the region. Iraq and Syria are also threatened by Iranian power ambitions, Iraq having already spent eight years on some of the bloodiest warfare in regional history against its Persian neighbor to the west.

Iran has also eclipsed other Arab states as the current main sponsor of global terrorism. Western intelligence analysts have dubbed its embassy in Bonn, Germany, the "nerve center" of Tehran's clandestine operations in Western Europe.

During a seven-year court trial in Germany, a former Iranian senior intelligence officer revealed the approval of assassination targets by Iran's Committee for Secret Operations. These "wet," or termination, directives were signed by Iranian leaders President Ali Akbar Hashemi Rafsanjani and the Ayatollah Ali Khamenei.

Along with Syria—which despite overtures of friendship to the West still remains one of the world's principal supporters of terrorism—Iran has sent secret cadres to help train various terrorist factions, including Palestinian Islamic Jihad, Sudanese militants, and the Japanese Red Army.

Iran maintains the largest navy in the Persian Gulf. The *Niruye darya-iye jumhuriye islami Iran (Nedaja)*, known to the West as the Islamic Republic of Iran Navy (IRIN), has a personnel roster of between 38,000 and 40,000 men, 18,000 of which are professional seamen.

IRIN operates approximately 110 vessels. These range in size from destroyers to fast attack craft and include submarines, hovercraft, and other naval vessels of all types. But numbers are not everything. The quality of IRIN forces doesn't match their quantity, with most surface vessels already obsolete, radars and other electronic warfare assets weak, and air defense flawed.

Because of these deficiencies and a lack of ready cash, Iran is upgrading its naval forces with an emphasis on getting the most bang for its military procurement buck. Therefore, one area of great interest to the mullahs who control the country is underwater warfare. Iran has purchased advanced diesel submarine technology from Russia in the form of three Project 877EKM boats. These are Kilo-class submarines designed by Russia's Rubin Central Design Bureau. The subs were shown at international arms shows in Europe and the Mideast, including the Satory '97 exhibition. Iran's Kilos are quiet and equipped with a sophisticated digital sonar suite enabling a counter-detection capability estimated to be five times better than that of Western boats of similar performance specifications.

The suite includes MGK-400EM sonar, MG-519 minehunting sonar, MG-512 propellor cavitation detection sonar and MRP-25EM search and warning radar. The Iranian Kilos are also equipped with advanced MVU-110EM combat information systems. These systems can track up to five simultaneous targets and give firing solutions on two targets concurrently.

Iran has also been investing heavily in other forms of antiship technology, including French Exocet and similar sea-skimming missiles, although its continuing acquisition of small, quiet, and electronically loaded subs poses the greatest threat to the United States and her Western allies. These nations must rely on force projection via sea power to the Mideast to achieve their military objectives and are thus especially vulnerable to undersea attack.

A report by the U.S. Navy released in 1997 calls submarines the "original stealth platform," and warns of the dangers posed by developing countries which can get their hands on custom-built and fitted boats from major arms suppliers. Among these, Russia is a recognized innovator, constantly upgrading the speed, silence, and stealth of their product.

Iran has also engaged in a project to develop nuclear warheads and ballistic missile delivery systems capable of striking theater targets. It has bought DF-15 and DF-11 theater ballistic missiles from China. Ironically, China may be repackaging U.S. missile technology for export because current U.S. pro-export policies enable dual-use technologies to flow into Beijing. This technology can be used to develop missiles and radar systems that China in turn exports to its trading clients.

Iran has also conducted test launches of medium range ballistic missile (MRBM) prototypes powered by the Russian RD-214 liquid-fuel propulsion system, originally designed for early Soviet SS-4 "Sandal" nuclear missiles.

Were Iran to successfully develop operational MRBM rocketry using SS-4 technology, it would have the capability of launching a roughly two-megaton thermonuclear warhead with a thousand-mile effective range. This means it could field a weapon capable of striking anywhere in Israel.

If MIRVing technology were used, then even a single strike could hit multiple targets. And with a small uptweak in combat range, Iranian MRBMs could be accurately deployed against Coalition warships operating in the Gulf, or even reach targets as far away as Southern Europe.

Although Russia had denied supplying missile technology to Iran, Boris Yeltsin reportedly signed an agreement with Tehran in 1996 to provide research and development assistance with next-generation surface-to-surface missiles. Ironically, it was the attempt by Russia to set up surface-to-surface missile launch sites in America's backyard in October 1962 that triggered the Cuban missile crisis.

Iran has also acquired Russian SA-10 "Grumble" air defense systems and has negotiated deals to acquire SA-10 surface-to-air missile (SAM) stockpiles to use in antiaircraft and anticruise missile defense. Iran's interest in building up its strategic defenses is linked to its efforts to increase its offensive capabilities: its leaders are concerned that the United States and/or her allies may launch preemptive strikes to take out its weapons production facilities and nuclear plants before its systems are perfected. The SA-10s are capable of hitting U-2 and SR-71 high-altitude reconnaissance aircraft fielded by the United States.

There are confirmed reports from the International Atomic Energy Agency and other sources that Russian and North Korean nuclear engineers were persuaded to live and work in Tehran via the promise of high salaries during the early part of this decade. Russian defense physicists driven from former government posts by the Soviet breakup were lured with incentives of $300 to $500 per "theoretical physics" problem solved. All the "problems" concerned components of nuclear weapons design, missile guidance, and other related weapons technology. Some of these solutions came into Iran's possession via the Internet.

Although denied by Tehran, U.S. intelligence agencies claim that Iran is engaged in a crash program to develop theater-capable nuclear arms. Since the Gulf

War (1990–1991), Iran has repeatedly been caught trying to evade U.N. embargoes against export of dual-use nuclear technology.

Two ongoing projects that harbinger this nuclear capability are the completion of the Bushehr nuclear power station on the Gulf Coast with Russian assistance and the construction of an installation at the Nuclear Technology Center in Isfahan to process uranium hexafluoride, a gas injected into ultra-high-speed centrifuges in order to produce weapons-grade uranium by "enriching" uranium isotopes.

Iran has also reportedly purchased surface-to-surface missile technology from the Ukraine, reputedly one of its principal black market suppliers of core weapons technologies.

Some of this illicit war materièl flows through Bolivia, a main conduit and transshipment center in major league black market arms trading, though lax enforcement sometimes means that even the United States and the United Kingdom have allowed arms to slip through the very U.N. embargo they created.

Trouble in the Transcaucasus: Armenia and the Turkey-Azerbaijan Axis

With the discovery of large oil reserves in the Caspian Sea, the Soviet breakaway state of Azerbaijan is now poised to become one of the world's largest petro-producers. By contrast, its neighbor Armenia, possessing few oil reserves and beset by internal problems, has entered an economic downslide.

A bone of contention between the two states is the region of Nagorno-Karabakh, over which the countries went to war and recently reached a shaky truce.

Each country is accusing the other of violating the peace treaty that ended the war. These allegations include charges of illegal weapons transfers to the combatants by both the CIS (Commonwealth of Independent States, also known as the FSU, or former Soviet Union) and Iran.

Both countries have also entered into strategic agreements with other countries to buttress their positions. Azerbaijan has forged growing military and economic ties with its northern neighbor Turkey, a formidable military power in the region and a key NATO member.

To counter the Turkey-Azerbaijan axis, Armenia has consolidated treaties with Greece and Bulgaria. It also has a much bigger and stronger friend in Russia, with whom it has had longstanding historical, cultural, and political ties. A sign of the developing relationship are two new Russian military bases on Armenian soil and joint military exercises by both countries.

Observers point to the Transcaucasus as a potential candidate for the next Serbia. The region is a strategic one to the CIS, for whom it represents a vital conduit for the flow of Caspian Sea oil. Russia cannot afford to have its flow of oil jeopardized by foreign control and might be motivated to support moves to depose the current Azerbaijani leaders and impose a heavy-handed "Pax Rus" in Nagorno-Karabakh.

Renewed fighting in this strategically sensitive area would likely draw in not only Turkey, but Iran as well. It could also open the door to U.S and European intervention to protect critical oil reserves needed by the Western powers.* Unlike the Gulf War, motivated by similar concerns, a war in the Transcaucasus would

---

*This was written prior to Allied Force, which seems to underscore my point.

pit Coalition forces against adversaries, backed by a
former nuclear superpower, who are far better armed,
far better trained, and far better motivated than Saddam's Gulf War–era Iraq.

The Pacific Rim:
The World's Wettest Tinderbox

Under a recent revision of 1978 guidelines for
U.S.–Japanese defense cooperation under the new
Treaty of Mutual Cooperation and Security signed at
the March 1997 summit between President Clinton
and Japanese Prime Minister Hashimoto, the Japanese
Self Defense Forces (SDF) now enjoy the greatest operational latitude since the close of World War Two.

Minesweeping, ship inspection, and intelligence
collection during an "emergency" in the Asia-Pacific
region are all mandated by the treaty. The guidelines
also call for the provision of "logistical support" to
U.S. forces by the Japanese.

In the past Japan has been restricted to supplying
U.S. land bases in the region only during emergencies.
But the catch-all term "logistical support" opens the
way to broad interpretation, including, at its broadest
reading, the provision of direct military support to the
United States in time of war.

North Korea is clearly nervous about Japan's growing military capabilities and its increasing military ties
with its major western ally, the United States. It is
especially nervous about the treaty's Article 5, which
states that America and Japan would act jointly
against armed attack on Japanese interests in the region. North Korea probably has genuine cause for
concern. The new guidelines appear to call for overt

Japanese military assistance to U.S. forces in operations conducted on the Korean peninsula. South Korea shares this uneasiness about the new guidelines, as does another strong regional contender, Beijing.

If there is a new spirit of belligerence wafting through the balmy Pacific air, it has not happened in a vacuum. The Pyongyang-based government of North Korea has been engaged in an effort to develop nuclear weapons, apparent since its 1994 test firing of the Nodong theater ballistic missile (TBM) whose range brought Japan within easy striking distance of Korean nuclear forces.

In January of 1997, Pyongyang announced a deal with Taipower, the Taiwanese state-owned power company, to transport 200,000 barrels of low-level nuclear waste to North Korea, supposedly for storage. Intelligence analysts believed that the so-called "waste" can and probably *will* be refined into weapons-grade plutonium as part of communist Korea's nuclear weapons development program.

China, too, has been a source of instability in the region, in part reacting against new and stronger alliances between its ideological adversaries. The new U.S.–Japanese security treaty is one of these alliances. There's also the 1996 Sydney Declaration, a U.S.–Australia treaty, an Australia-Indonesia security agreement, and the incorporation of Vietnam and India into ASEAN (with Burma and Laos expected soon to join), a Pacific version of NATO.

These alliances threaten Beijing's push for supremacy in the region and its plans to become a global economic and military superpower in the twenty-first century. Its recent naval maneuvers in the South China Sea and the Indian Ocean show that China has no intention of being pushed into a corner by what it perceives as a new pro-Japan and pro-Western regional alignment.

On the flipside, the historical militarism and imperialistic aims of Japan might also prove to be the matches that ignite the powderkeg of tomorrow's naval war in the Pacific. In recent years, there has been popular support in Japan for the nation to regain something of its former imperial glory.

With an enhanced role for the SDF, Japan could find itself tempted to overstep the bounds of the treaty and engage in something like the military adventurism of its World War Two expansionist days. In this case the United States might find itself drawn into the fight—this time, not without irony, as an ally of the once-demonized bombers of Pearl Harbor.

A strategic scenario that might see U.S. troops involved in fighting in the Pacific Rim might involve the invasion of South Korea by its northern communist neighbor, perhaps as a preemptive measure of some kind, which would almost surely bring U.S. troops to the Pacific Rim.

Likewise, a clash between Japan and the communist North Korean state or China could lead to regional warfare and subsequent U.S. military involvement. In either case, a combination of U.S. naval engagements and aerial bombardment of strategic targets would quickly ensue, with ground forces used only if a cease-fire were not quickly put in force.

But it is entirely likely that before anything like this happened, the situation already could have escalated to the point of nuclear war.

OOTW and MOBA: Operations Other Than War and Military Operations in Built-up Areas

As their names imply, OOTW and MOBA operations encompass everything short of conventional conflict. OOTW runs the gamut from peacekeeping efforts, such as recent U.S. interventions in Haiti and Somalia, to counterinsurgency and counterterrorism missions, including counterdrug operations.

It has also come to include operations directed toward stemming tides of illegal immigration or mass migration. This is being viewed increasingly as a potential threat to national security by many countries including the United States which detained and forcibly repatriated waves of Haitian and Cuban refugees in 1994 and Chinese boat people in 1995 and 1996. Fleeing Kosovars in 1999 are the latest on this sad list.

Critics have charged that OOTW turns military forces into a global police force, and Uncle Sam into a big cop walking a planetary beat. But there are always valid humanitarian concerns regarding the use of military troops in refugee control missions, and if recent history is any kind of a guideline, the likelihood that tomorrow's soldier will find himself involved in some form of OOTW operation is among the highest of all possible scenarios. And as we have seen recently, operations other than war can turn nasty at the drop of a feather, especially in urban locales where MOBA scenarios are likely.

During what began as peacekeeping roles in Haiti and Somalia, United States ground forces discovered that to become enmeshed in a violent conflict in an urban environment having its roots in complex, ancient, and interwoven grievances put them quickly and unalterably in harm's way. More recently, U.S. observers of the uneasy truce in Serbia have found themselves taking fire of the most unfriendly sort from factions supposedly no longer at war.

By the same token, interdiction missions undertaken by elite commando-style units against terrorist, narcoterrorist, or guerilla forces seldom come off without fire being traded, even if everything goes down like Teflon-coated clockwork.

While actual operations of this type by U.S. SEALs, Delta Force, or other elite SOF (special operations forces) units have never "officially" taken place, two commando assaults undertaken by the French GIGN *(Groupement d'Intervention de la Gendarmerie Nationale,* colloquially called *"Gigène"* and the British SAS (Special Air Service) can be studied as cases in point.

In the first, French commandos staged a lightning assault and hostage rescue operation on a grounded Air France 747 at Paris's Orly Airport in 1995. Although the operation cleared the aircraft of the Algerian terrorists who'd hijacked it, there were casualties.

In the second case, British commandos freed hostages held at the Iranian embassy in Palace Gate, London, during a terrorist siege in 1980. Here, too, friendly casualties occurred. In both instances, the violence was brief but brutal.

Ongoing development of nonlethal weapons, such as infrasonic nerve agents, low energy lasers, ammunition that disables but is not fatal, and incapacitating chemical sprays of various types is intended to reduce the level of violence and pursue a soft-kill policy of deterrence. Moreover, improved surveillance and intelligence processing technologies mean that military strategists will be better able to plan missions than in the past. Better training methods will play their part, too. All the same, there is no guarantee that tomorrow's soldier will not find himself placed squarely in harm's way during even the most minor engagement.

As we have seen, the post–Cold War peace is shaky terrain sown with dragon's teeth, any one of which could sprout up into a future regional conflict. The road to tomorrow's battlefield could be an express lane we may soon find ourselves traveling.

THREE

• • • • • • • • •

War Zones and Weaponry 2010

THERE HAVE BEEN many battlefields throughout the violent history of warfare. Their names stand out as bloodstained milestones in mankind's tortured path toward an uncertain future.

From Marathon to Agincourt, from Waterloo to Anzio, and from Vietnam to the Persian Gulf, these killing grounds have shared at least two things in common.

First, they have frequently changed the course of history. Second, innovations in combat technology have often been deciding factors, bringing glory to the victor and destruction to the vanquished.

It's fair to predict that the coming combined arms battlespace of distributed operations will share these two familiar aspects of all warfare. Yet tomorrow's war zone will be radically different from all that have gone before it.

Even much about the Gulf War's prosecution is now outdated. The AirLand battle doctrine that drove Coalition forces' order of battle has been superseded by other doctrines designed to exploit new technologies and warfighting concepts.

The weaponry of the Gulf, largely drawn from armament stockpiles built up since the end of the war in Vietnam and utilizing Cold War-era technology, is being replaced with smarter and deadlier systems. Even the grunt on the ground is in the process of undergoing radical changes.

The battlespace of the period 2010 to 2020 will be a combined arms battlespace. Information technology (IT) and stealth will be the prime enablers toward dominance by friendly forces. Jointness and interoperability of all forces and combat systems will be the cornerstone of doctrine.

Communications will be digital and networked. Weapons systems will be modular.

Combat operations will be nonlinear, taking place in multiple dimensions, extending beyond physical boundaries into the electromagnetic spectrum, orbital space, and even the minds of the enemy.

"Bang 'em and burn 'em" will still be the order of the day, but on an entirely new level of attrition.

Following the end of the Cold War, the U.S. Department of Defense (DOD) conducted a "bottom-up review" (BUR) of United States military forces.

The BUR was designed to reassess force structures, redefine doctrine, and set new goals in light of an emerging New World Order taking shape in the aftermath of the collapse of the Soviet empire and the disintegration of the world of bipolar superpower conflict that had driven U.S. foreign policy since the end of World War Two.

From the BUR emerged a military force over 33 percent slimmer and tasked with a new and historic mission. In place of readiness to fight a European land war against the Soviets and their Warsaw Pact allies and a second "half-war" somewhere else in the world, America's warriors were tasked with preparedness to fight two major regional conflicts (MRC) and engage in simultaneous OOTW and/or MOBA efforts.

The new, leaner force would leverage U.S. high-technology assets to compensate for its smaller size. It would be leaner, yet it would also be far meaner. It would be a scalpel instead of a mace, wielded by commanders trained to field battlefield assets with an almost surgical sense of precision.

The newly completed Quadrennial Defense Review (QDR) is the nation's second bottom-up review and marks the start of regular four-year reviews of the U.S. military force structure and weapons systems.

As congressional observers had expected, the QDR continued the scaling back of military programs and further reductions in military force levels that have marked the post–Cold War years. At the same time, the review gave the Pentagon a clear mandate to continue the revolution in technology, tactics, and training that has been molding the U.S. armed forces into the superforce of the twenty-first century.

Joint Vision 2010 is the plan formulated by the U.S. Joint Chiefs of Staff to implement the development of tomorrow's military force. In turn, the three armed service branches have all developed their own integral plans in response to JV 2010's guidelines.

The cornerstones of JV 2010 are four dynamic concepts: full spectrum dominance, precision engagement, full dimensional protection, and focused logistics.

Taken together, these concepts call for strategies to control and shape future battlespace, advantageously engage opposition forces by lethal and nonlethal means, maintain the integrity of deployed forces, and control the tempo of operations (OPTEMPO) to bring about a swift and decisive victory.

From JV 2010 each service branch has prepared its own strategic plan for warfighting in the next century. Force XXI is the Army's plan for land combat in the coming years. Global Engagement is the Air Force's concept and Copernicus Forward or N6 is the Navy's version of JV 2010 implementation. Let's now take a closer look at what these plans entail.

Force XXI:
Land Combat in the Twenty-First Century

"The Information Age is upon us. We must change in order to sustain current levels of excellence into the future. The very nature of warfare is changing and our army is fully engaged in an effort to determine how we will operate in this new environment," reads the introduction to a white paper issued by the U.S. Army Training and Doctrine (TRADOC) Command based at Fort Monroe, Virginia.

Combined arms force integration, digitization, multidimensional battle tactics, advanced weaponry, and enhanced command, control, communications, computing, and intelligence (C4I), global power projection, and information dominance are all part of the U.S. Army's vision of Force XXI combat operations.

These encompass several key points:

Force XXI operations are multidimensional. The battlespace in which they take place exceeds traditional dimensions of height, width, and depth.

Global digital communications extend situational awareness from the combat zone to orbital space and back to the home front while the invisible hands of electronic warfare (EW) and information warfare (IW) attack the enemy's C4I capability.

Mission, enemy, terrain, troops available, and time (METT-T) is a concept important to the understanding of the new doctrinal thinking. METT-T drives and shapes the battlespace.

METT-T determines the application of firepower and the deployment of combat personnel. METT-T determines the allocation of intelligence, deception and psychological operations resources, and the placement of military infrastructure. METT-T, in short, is the operational linchpin around which all else pivots.

Force XXI operations are surgical. Attacks are synchronized, planned with precision, and executed with violent fury to strike at the Clausewitzian "centers of gravity" in order to deliver a knockout blow early on in the engagement.

Advanced networked digital communications, field sensors, and simulations enable commanders to plan, rehearse, and execute operations to maximize the force of the attack.

Force XXI operations are nonlinear and distributed. They take place simultaneously across the entire battlespace rather than massing troops and hardware at a single point, as in old-style tactics. They are something like a blitzkrieg in many dimensions.

Driven by METT-T, these operations are fluid. They change and metamorphosize throughout the duration of the campaign to meet the changing demands and situations encountered during the prosecution of the war. Force is concentrated at decisive points in response to changing tactical developments, not statically applied according to some preconceived plan.

Force XXI operations are integrated into the combined arms doctrine. Jointness and multinational integration are basic tenets of the program.

Army combat elements will be designed to function as part of a joint task force comprising Coalition partners and operating in conjunction with nongovernmental agencies such as the United Nations. These elements will share physical resources and intelligence, and will engage in joint missions with elements of other armed services in the Coalition, both national and transnational.

Various global combat operations and advanced warfighting experiments (AWE) conducted by the U.S. Army have validated many of the Force XXI concepts outlined above.

Desert Hammer is one of these AWEs. It put Tactical Force 1-70—a battalion-sized combined arms combat group equipped with advanced networked digital communications equipment—into a two-week live-fire, force-on-force training exercise in a desert environment that presaged a future Mideast war.

The results of the AWE showed that virtual reality simulations technology combined with networked digital communications, those which combined voice with visual imagery, enhanced the ability of the troops to better make sense of what was going on around them, and so to better carry on the fight.

Another AWE, Prairie Warrior, deployed a division-sized mobile strike force equipped with advanced digital fire control systems in simulated engagements against a mock enemy force in a land battle environment resembling that which might be found in a future European battle theater.

This warfighting experiment demonstrated that massive and pinpoint-accurate firepower followed up by high-speed maneuver warfare made possible by advanced networked information systems quickly smashed enemy defenses with minimal friendly casualties.

Finally, from global OOTW and MOBA operations, such as Restore Hope, Uphold Democracy, Able Sentry, and Joint Endeavor, the Army has gained other insights into Force XXI concepts.

In Able Sentry and Joint Endeavor, U.S. peace-keeping troops in Bosnia have used networked digital communications systems, unmanned aerial vehicles (UAVs), robotic countermine vehicles, and computer-based translingual communications devices in the field. In relief operations in Somalia, Restore Hope forces used a sophisticated tracking system (TAV) to monitor the flow and distribution of supplies and resources.

It remains to be seen whether toughness and the will to fight and die—the qualities that generals and foot soldiers alike throughout history have called the key to winning land wars—measure up against technological superiority and agile combat doctrine. Tomorrow's enemy may have the one, whereas we may possess only the other. But as to which vision will prevail when, to paraphrase Mao Tse Tung, politics again "turns bloody," only the future will tell.

Global Engagement: Air and Space Deployment

The U.S. Air Force (USAF) has also redefined its priorities in the wake of post–Cold War realities. Global Engagement is a forward-looking strategic policy following on from Global Reach-Global Power, the USAF's vision statement during the transitional period subsequent to the Cold War.

Global Engagement's cornerstones are the development of four primary capabilities to create air and space dominance for the USAF. The first cornerstone is to strike directly at an adversary's centers of gravity while simultaneously hitting its operational and tactical warfare infrastructure.

The second cornerstone is global situational awareness—the ability to link real-time data from air combat elements to command centers and political centers back home through high speed networked digital technology so that military operations can be seamlessly orchestrated across the entire operational theater.

The third cornerstone is the ability to bring intense firepower to bear over global distances within a matter of hours or days at the outside in order to leverage the military resources of the U.S. and her Coalition allies. Stealth, speed, and precision-guided munitions contribute to this drive.

The USAF is also now engaged in transitioning from an air force to a space and air force, Global Engagement's fourth cornerstone. Space is linked to military operations on land, at sea, and in the air. At the same time, key military functions are conversely migrating to space.

Surveillance and reconnaissance operations are routinely conducted by military and spy satellites, such as Lacrosse, Keyhole, and Magnum, orbiting high above the planet. Satellites are also critical to the development and maintenance of high-speed digital military communications networks and global positioning systems.

Without assets in space, a high-technology force such as the U.S. military would be unable to see, hear, or think. Instead of stealthy ninja warfare, the name of the game would be the kind of bloody, grinding,

force-on-force engagements that have marked the An-
zios, Salernos, Paschendales, and Chechenyas of military history.

Without space, in short, there could be no such
thing as a digital or combined arms battlefield.

As the envelope of military operations continues to
expand, it will encompass space more and more. And
as other nations begin to develop offensive space capabilities, they will attempt to strike at U.S. and Coalition space assets in any opening engagement to a
future war.

Wargames separately conducted by the Army's
Task Force 21, the Strategic Command, and the non-
military U.S. Space Foundation's Space Symposium
have all shown that "attacker" forces see U.S. defense satellites as the jugular vein of our military
power and immediately try to cut it upon commence-
ment of hostilities.

There are many options for this form of attack. All
it takes to wreak widespread havoc is the detonation
of a low-yield nuclear bomb in low orbital space.
The resulting nuclear-electromagnetic pulse (N-EMP)
blinds satellite sensors and burns out sensitive electronic components.

Almost instantly, communications nodes blink out,
stealth fighters lose navigational control, naval carrier
groups steam around in circles. This style of attack
could be termed the "Murder Incorporated" gambit,
for it is the global equivalent of sticking an icepick
into a human brain.

As this book goes to press, the Scud missiles available to Third World dictatorships do not possess anything near the range necessary to strike at U.S. space
assets. For the immediate present, at any rate, rogue

states might as well throw pebbles at the moon as attack U.S. space assets.

Apart from our Western allies and possibly the Israelis, only the Russians and the Chinese possess viable antisatellite technology, and they, for the moment, are unlikely to make use of it for various sound strategic and political reasons, such as being blown up by Tomahawk missiles in a retaliatory strike. But by tomorrow's war, Iraq, Angola, the Sudan, Syria, India, and Pakistan will almost certainly all have that capability. Those pebbles will have turned into meteors.

While long-distance unmanned aerial vehicles and manned high-altitude reconnaissance aircraft can provide backup resources in the event of the destruction of space-based military assets, they are no true substitutes for satellites.

And while satellites can be armored and equipped with countermeasures, the only viable form of protection is the development of a true space force comprising military space stations and air-to-space capable planes using advanced scramjet engines that are air-breathing rocket hybrids. As these developments put us somewhat beyond tomorrow's war, we'll discuss them later on, when we consider what the more distant future may hold for warfare.

Copernicus: Exploiting the Battlecube

The USN (U.S. Navy) implementation of JV 2010 is called Copernicus. It is named for the scientist-philosopher whose radical thinking formed the basis of modern astronomy. The symbolism is apparent—

the USN's goal is no less than a radical break with past naval combat doctrine.

Like the Army and Air Force programs, Copernicus has developed a strategy to put systems in place that use digital information processing computer nets, space-based assets, and high technology to create the global naval force of the twenty-first century. The USN calls this its "Forward from the Sea" (often shortened to "From the Sea") strategy, and the various strategies developed in support of Copernicus programs "architectures."

The thing to bear in mind about Copernicus is that it's essentially about information and tactics, as well as the systems to implement them, not so much about the acquisition of major new combat hardware.

With the exception of naval variants of the planned Joint Strike Fighter, the stealthy Sea Shadow program, and upgrades to existing components of the U.S. fleet, such as nuclear carriers and Aegis cruisers, the USN has few big-ticket acquisitions programs in place for the near term.

Because there's little in the way of new subs and ships on the drawing board, some critics have labeled Copernicus a mere "bumper sticker." The Navy thinks differently.

Bumper sticker or not, the USN approach to tomorrow's warfare is markedly different from that of the USAF and Army. It is especially so vis-à-vis the case of the USAF, which has major hardware acquisitions planned, such as the F-22, the B-2, and advanced unmanned aircraft of various types. Information flow, management, and exploitation in combat as represented by C4I—command, control, communications, computing, and intelligence—forms a high-speed information envelope surrounding deployed naval forces in theater.

This is the "battlecube" that Copernicus wants to operate within.

Copernicus has several cornerstones. One is to create a next-century naval expeditionary force with global reach, one that can rapidly project U.S. military punch to trouble spots around the world at high velocity and with enough impact to extinguish global flare-ups at short notice.

As mentioned earlier, sea power will become America's most vital military priority in the twenty-first century as the United States becomes the world's greatest maritime power and veers into almost inevitable conflict with other seafaring nations such as China who would control vital portions of the world's oceans.

The USN will also play an expanded role in the combined arms battlespace of the twenty-first century.

Sea power has in the past been used primarily to conduct battles on the open seas or to provide limited support for ground forces in coastal areas. Such a "blue water" naval capability is being transitioned to a navy force structure optimized for two kinds of warfare—"littoral" (in and around the world's coastlines) and "expeditionary" (projecting global military power).

It's in coastal areas that the world's major population densities now lie, and where the USN will need to "kick in the door" to provide viable support for friendly air and ground attack formations.

The Navy will do some heavy hitting of its own, too. Of all the images of the Gulf War, one of the most memorable has been the startling performance of Tomahawk cruise missiles launched not only from

U.S. surface vessels but from submerged ballistic sub-marines as well.

In the age of the cruise missile, naval forces have come to play a major role in land engagements. This is a role that will continue to grow and develop as tomorrow's wars approach.

Naval aviation will also continue to play an expanded role in future military actions, not only to protect carrier battle groups against aerial attack, as it has traditionally done, but also to project military power toward land-based and space-based targets.

Naval versions of advanced fighter aircraft will multiply the striking range and firepower of sea-based vessels by several orders of magnitude. The old salt's adage that "anywhere you can carry a canteen" is where the Navy has a mandate to fight will have become a part of standard military doctrine.

For these reasons, the USN will need to become the most "joint" of all the armed service branches in an era where jointness is the first commandment of military operations.

It will need to put in place strategies, doctrines, and technologies that will enable it to interact with such diverse tactical information assets as AWACS and JSTARS aircraft, land-based C4I centers, and the naval forces of Coalition partners. It will need what military planners call a "system of systems."

In pursuit of these Copernicus architectures, the USN is developing a comprehensive sensor-to-shooter battlecube environment made possible by two systems. The Joint Maritime Combat Information System (JMCIS) and the Global Command and Control System (GCCS) will drive innovation for the warfighter.

Recent Fleet Battle Experiments (FBE) such as

FBE Alpha and Bravo that have tested proposed naval force structures in pursuit of the "From the Sea" strategy, have already proved the value of these systems. But by exploiting the battlecube, the Navy is pursuing the most information-based strategy of all military arms, and one that may need considerable development to live up to its full potential.

DARPA: Tomorrow's Weapons Today

The Defense Advanced Research Projects Agency (DARPA) is the little engine that drives the big technology that America will depend upon to create the combined arms force of the twenty-first century. DARPA's small size is deliberate, not accidental.

On his appointment, new director Larry Lynn took an ax to the nascent bureaucracy that was beginning to take form at the agency, seen by him as a "fatal affliction." Lynn, a former deputy director in the 1980s, envisioned DARPA as lean, mean, and clean, its small size and budget encouraging experimentation and improvisation critical to cutting-edge R&D.

DARPA's mandate is to focus on high-risk, high-payoff technologies and military concepts. First there are the quickies—projects that have limited lifespans and are continuously reviewed to weed out those that are dead ends or low-end producers.

Then there are the high-end programs—critical military systems and technologies that will need time to mature and whose development cycles can run to years and decades. Stealth is a good example of these.

Last but not least, DARPA is the real-life version of the fictitious Bondian Q-Branch—when problems arise, the agency tries for a technological quick-fix.

When the situation calls for an exotic or offbeat tool to do the mission, DARPA aims to develop one rapidly.

DARPA is like the fictitious Q-Branch in another way, too: it's the main clearinghouse for covert "black world" programs that coexist with open or "white world" programs.

Sometimes these black programs are entirely separate from the types of concepts DARPA is working on in its open research and development. Most of the time, though, white programs develop technologies that are transitioned to secret black programs, and vice versa.

Often, a white program will be canceled or be shut down due to "lack of funding" or because "political developments no longer warrant it" only to continue in its clandestine development mode, very much alive and producing arcane technologies that will form the basis for tomorrow's secret weapons. One of the reasons such programs go underground is because even acknowledging the existence of a revolutionary new technology with important military applications can give adversaries clues to how to mimic or defeat it.

Stealth technology, now openly acknowledged, though for years one of DARPA's most closely guarded secrets, is one example. A more current example may be the existence of hypersonic aviation technologies based on years of DARPA research, the so-called Aurora Project.

It's hypothesized that Aurora aircraft—manned and unmanned space-capable planes—exist and are capable of flying at better than five times the speed of sound. It is believed that these are already being used to perform high-altitude surveillance missions too risky to be undertaken by other spy planes.

Overt or covert, DARPA's research and development efforts usually result in ACTDs or Advanced Concept Technology Demonstrations. ACTDs converge near-term technologies with existing military weapons systems that can be prototyped and fielded for early evaluation on short notice. This provides room for feedback from combat branches that carries over into the development cycle, speeding it and enhancing it.

Looking ahead to the combined arms environment foreseen by JV 2010, DARPA is focusing its development efforts on enabling technologies, such as information technology (IT), that will help give the United States its leading edge in future conflicts.

Lynn sees new-age threats from biological attack, cruise missiles, and information war and new enemies in the form of rogue states and terrorist groups as potential twenty-first century adversaries.

DARPA's ongoing programs reflect this worldview and are tailored to provide solutions to these threats. Some of DARPA's priorities are in the areas of defensive information warfare, high-bandwidth/high-definition communication systems, smart materials and structures, advanced training simulation, and battlefield medicine.

Robotic Reconnaissance and the Electronic Order of Battle (EOB)

The order of battle in tomorrow's combined arms battlefield will include planes, tanks, ships, and many other elements of a military force structure that have become commonplace and familiar to us.

But warzone 2010 will also have another order of

battle, an electronic order of battle (EOB). The EOB will run alongside the physical order of battle. Its province will be the electromagnetic spectrum, a realm of microwave emissions, infrared radiation, and the digital world of cyberspace.

Part of the EOB will be devoted to the conduct of offensive and defensive information warfare (IW) which encompasses the somewhat different sphere of electronic warfare (EW).

EW specifically describes the application of electromagnetic fields to jam, spoof, surveil, or destroy enemy command, control, and communications systems and to counter unfriendly attempts to do the same to friendly C3I systems.

IW goes beyond EW to include digital information processing systems such as computers, computer networks, and the software that these use, and even includes by some definitions the "wetware" of the human brain and nervous system, which, by exposure to sonic, light, and other nonstandard weapons, can be affected in ways similar to that by which digital viruses affect the information processing capabilities of computers.

Robotic devices of various kinds will figure strongly in the EOB. These will roll, crawl, or bestride the land, fly through the atmosphere and the reaches of near-space, prowl the oceans and the littorals, and burrow beneath the ground.

They will go to places and do things that would be deemed too dangerous for the human soldier to undertake. Apart from being natural weapons platforms, the robots of tomorrow's war zone will find ample employment as intelligence collection assets.

Unmanned aerial vehicles (UAV) and unmanned underwater vehicles (UUV) are robotic flying and

swimming machines. The various types under development today will become important players in tomorrow's theater and strategic warfare.

Small, stealthy, silent, and smart, they are capable
of being crammed with gear that can collect and dis-
seminate real-time battlefield intelligence to remote
command and control centers. Some of these robotics
will be hummingbird-sized and could be carried along
with an infantry soldier's gear. Others will in fact not
be small at all, but may be as large as manned aircraft,
although completely automated.

Some UAVs will act as smart, agile decoys for advanced combat aircraft while UUVs will perform
much the same task for ships and submarines. Other
battlefield robots will find their principal uses as mechanical moles or ferrets that sneak into underground
installations and report back on what they have found.

Other such systems will be land cruisers, rolling
through the night on deep-penetration spy missions or
positioning themselves to launch missiles covertly at
enemy targets both human and mechanical. Still other
types of robot vehicles will cruise through orbital
space protecting friendly satellites and destroying the
enemy's offensive orbital capabilities before they can
strike.

Tier III Minus and Tier II Plus

At 2:25 A.M. on March 29, 1996, a small unmanned
robot plane with a discoid body and two long, narrow
wings jutting out at its rear took off from a remote
desert airstrip at Edwards Air Force Base in Northern
California. The airstrip was at the Air Force Flight
Test Center, a facility that has for several decades

been no stranger to an array of arcane flying devices.

The UAV's name was Darkstar, officially known as Tier III Minus and developed under the High Altitude Endurance Unmanned Aerial Vehicle program (HAE UAV) managed by DARPA for the Defense Airborne Reconnaissance Office (DARO).

Darkstar is characterized by DARPA as a "high-altitude, endurance unmanned air vehicle optimized for reconnaissance in highly defended areas." The Darkstar UAV flew for 20 minutes, during which it reached an altitude of 5,000 feet and executed a series of preprogrammed flight maneuvers. It then returned to the launch site and landed. Darkstar did everything autonomously, acting completely on its own.

When it becomes fully operational, Darkstar will be able to operate out to a range of 500 nautical miles from its launch site. It will be able to loiter over its target area for over 8 hours at an altitude of 45,000 feet while its electro-optical (EO) or synthetic aperture radar (SAR) imaging payloads (the UAV will not carry both simultaneously) transmits high-resolution graphics of whatever it's been sent out to spy on far below in near-real time (NRT) over either satellite communications (SATCOM) relay or line-of-sight (LOS) communications links.

Darkstar's big brother is Tier II Plus, a thirteen-ton single-jet engine aircraft with a 116-foot wingspan otherwise known as Global Hawk. The size of a single-engine plane, Global Hawk is a sleek drone aircraft whose fuselage seamlessly blends into a huge ramjet intake at the rear and flares into a rounded sensor bulge at the front, with long, slightly swept glider wings sprouting from its sides. The combination of design elements gives it the look of a winged

monster blindworm full of cold, sinister intelligence—
which is not very far from what it is.

Global Hawk is intended to fly "long endurance"
missions that require it to stay airborne—or, as it's
said in military parlance, "loiter"—for as much as
25 hours before returning to base. During that time it
will be able to aerially survey approximately 40,000
square nautical miles, a target area equivalent in size
to the state of Illinois, with a resolution of objects as
small as three feet in diameter (spot-mode scanning
will enable it to surveil a much smaller area with a
resolution of objects down to one foot in size).

Like Darkstar, it will also be equipped to carry EO
and SAR imaging payloads. Unlike Darkstar, however, it will be able to carry both systems and a third
infrared (IR) system simultaneously while operating
at altitudes of greater than 60,000 feet—the UAV can
be crammed with up to a ton of cameras, radars, and
other sensors. Global Hawk will also transmit NRT
data via LOS datalink and SATCOM relay, although
these channels will enable transmissions of higher resolution than will Darkstar.

A variant of Global Hawk might also be built as an
armed drone equipped with weapons to enable it to
shoot Scud missiles out of the air. In pursuit of this
objective, a joint U.S.-Israeli program called MOAB
uses a kinetic energy (KE) weapon capable of firing
projectiles at ultra-high speeds at ballistic theater missiles while in their boost phases.

This is similar to what the Patriot missile defense
system does, although it uses a different approach.
Unlike Patriot missiles, the KE projectiles are not explosive warheads per se. Their extreme velocity
causes them to transfer tremendous amounts of energy
to the target, which, if struck, will be broken up in

flight. Approximately $31 million will have been spent on this project through fiscal year 1998. MOAB-armed Global Hawk UAVs could also theoretically be used to kill other UAVs, enemy aircraft, and even satellites in low earth orbit.

While Darkstar and Global Hawk are large un-manned planes, far smaller UAVs are also being developed. At the other end of the UAV size-spectrum are MALD and MAV.

MALD stands for miniature air-launched decoy. MALDs would be carried onboard fighter and other aircraft and released when enemy radars threatened to acquire friendly aircraft as targets. The MALDs would then use their onboard computer systems to jam and spoof the radars, creating false echoes and other effects while the actual plane executed high-speed evasive maneuvers.*

Even smaller than MALDs would be MAVs, or micro-air vehicles. The extremely small size of MAVs will optimize them to perform military missions that larger drones could not accomplish because of stealth, maneuverability, or other issues associated with their size.

MAVs the size of hummingbirds might be sent out to detect biochemical agents in contaminated areas, act as tiny communications satellites or radar decoys, be used as nasty little flying antipersonnel or antiar-mor mines, and also fly down airshafts for peeks into buildings in urban warfare environments.

The Distant Portrait (DP) program run by DARPA

---

* MALD flew for the first time on January 9, 1999 in a test at the USAF Flight Test Center at Edwards Air Force Base, where it was successfully launched from an F-16 fighter plane and recovered intact.

is looking into the use of MAVs in this last-named role, but is also developing creeping, slithering, and rolling land-based robotics that are very tiny and very stealthy and can probe DUFs (deep underground facilities).

Distant Portrait probes will scout for NBC (nuclear-biological-chemical) weapons or materials and transmit the data back to base (which will most likely be a covert team stationed nearby with a remote control unit instead of a static command post), then either self-destruct or return to sender, there to await another go-round.

Other robotic land systems to be used on tomorrow's battlefield will incorporate hybrid electric high-mobility vehicles with both intelligence-gathering packages and weapons delivery systems onboard. Like systems used in Bosnia today, they will also perform countermine activities.

A joint Army–Marine Corps operation requirement document (ORD) issued in April 1997 called for a tactical unmanned vehicle (TUV) system capable of conducting remote reconnaissance, surveillance, and target acquisition (RSTA) and carrying out chemical warfare agent detection missions. The TUV would be controlled by a robotic OCU (operator control unit) and would have a modular design capable of being quickly changed or upgraded to accept new technologies and payloads, including weapons payloads.

A kindred concept which is also the subject of ongoing research and development at DARPA is the reconnaissance, surveillance, and targeting (RST) vehicle. The RST concept differs from the TUV principally in that, unlike the TUV, an RST would also be capable of carrying special forces personnel when configured for human carriage.

The RST would be a hybrid electric powered vehicle, using a mixture of electric and internal combustion motor technologies and using a beefed-up suspension system enabling it to move quickly through hard terrain or bad roads. It would also be low-observable.

Both these designs mesh perfectly with another DARPA program, the advanced fire support (AFS) system, which is intended to develop robotic missile and gun systems that can automatically deploy themselves, arm themselves, acquire their targets, and autonomously and stealthily attack these targets according to a set of preprogrammed rules of engagement. AFS systems are poised to take part in advanced warfighting experiments in fiscal year 1999.

The U.S. Navy, while fielding programs to use UAV technologies, such as recent programs to use the Darkstar UAV in conjunction with nuclear submarines, is engaged in development efforts for unmanned underwater vehicles as well.

Apart from aerial reconnaissance missions, the USN especially needs combat systems that can give its deployed forces early warning of deadly mines. Mine technologies are improving all the time and have become extremely sophisticated, capable not only of lying dormant for long periods of time and detonating when specific types of ships and subs approach, but also of launching torpedos from standoff range.

The long-term mine reconnaissance system (LMRS—pronounced "Elmers") is one of the UUV prototypes developed by the Naval Underwater Warfare Center (NUWC) which are based on tethered and untethered designs—that is, which can be guided by umbilicals or carry on their work autonomously, like their airborne counterparts.

LMRS will be a follow-on program to NMRS, the near-term mine reconnaissance system which is currently being fielded by the USN and is a less sophisticated system providing an interim solution to mine countermeasures and detection while LMRS is being developed, prototyped, and tested.

LMRS would be capable of operating both in open ocean or "blue water" environments and in shallow inshore or "littoral" environments. It would map and classify mines it found to enable navigation around them or in situ disarmament if all else failed.

A far more radical UUV program is the Manta project, also under development by NUWC. Still in the concept stage, Manta may be ready for naval operations by the year 2010.

Mantas would attach to submarines similar to the way suckerfish attach themselves to predatory sharks. Future submarines would be equipped with four or more Mantas. The Manta cluster would be embedded conformally in the submarine's hull, seamlessly blending with the hull contours, and made of stealthy composite materials to defeat sonar identification.

While attached to the exterior of the sub, the warfare systems onboard each Manta would function as part of the submarine's weapons and sensor array, under the control of the commander. But when these independently capable vehicles were launched, each Manta could be controlled separately and remotely.

The Manta's design, sensor suite, and weapons configuration are all as radical as the system's imaginative core concept. The wedgelike Manta is shaped something like a surfboard, thin and blunt-nosed in front and flaring toward the aft section, which is bent downward into two finlike stabilization surfaces and which supports an aircraft-style rudder. The rudder carries conformal communications antennas and upward-staring optical systems.

Sonar, ocean environment, and acoustic and non-acoustic sensors are located conformally at the Manta's nose, while other conformal arrays take the form of strips along its flanks. The envisioned propulsion system would use hydro-thrust rather than propeller action, helping to give the Manta a very low acoustic signature.

Inside the Manta's composite-material hull would be located an array of weapons systems including directed energy weapons such as particle beam guns and hyperkinetic-hypervelocity weapons. These would eject through a series of front and rear-facing tubes of various sizes, firing right through ovoid slits in the hull. A "short-range self-defense weapon," presumably some form of undersea missile, would also be carried internally and be fired from the nose.

A small-scale concept Manta demonstrator is due by mid-2000, if DARPA agrees to fund the Manta program for the Navy, to be followed by a full-size and fully operational prototype by 2002. The concept demonstrator would bear a $5 million price tag.

Except for Manta, all the above programs are currently under development and most have already produced fieldable prototype systems that are on track for deployment in the next major regional conflict. Between now and then, more types of military robots and more sophisticated robotic warfare systems will unquestionably be built to meet emerging threats.

Robots will take their place in tomorrow's battlefield in increasingly greater numbers, replacing the soldier-in-the-loop and automating a wide assortment of combat activities formerly performed by human beings.

In fact, the day may be foreseen when battle will be automated to the point where most combat platforms are unmanned and the outcome of warfare, of victory and defeat, will be based not on the traditional body count of yesterday, but on the number of intelligent war machines killed. But that day is probably still a long way off, and for the immediate future, the gods of war will continue to demand their customary blood offering.

Low Intensity Conflict and Special Operations Forces

Neither General Dwight D. Eisenhower nor General George Marshall, the two men most responsible for prosecuting the invasion of Europe and the war against Germany in the mid-1940s, had much use for the unconventional side of warfare.

This was the era in which the concept of the "citizen soldier," first pioneered by the armies of the French Revolution in the eighteenth century, was an unshakable cornerstone of U.S. military doctrine, and elite units were openly disparaged.

Such was not the case with America's allies or enemies during World War Two, nor, for that matter, the situation on the ground, even among U.S. forces, in several instances. Both the British and the Germans made use of special operations forces (SOF) as surgical tools when the need arose. They realized that the unconventional side of war—its dark underbelly, so to speak—was as important as sanctioned engagements on the level of battalions, division, and corps.

It has also been documented that American G.I.s carried out commando-style raids on enemy positions during the stalemated fighting in the Aurunci mountains of Italy during the winter of 1944. Many of these G.I.s belonged to the 36th Infantry Division, the so-called T-Patchers, or "Texas Infantry," who possessed a guerilla warfare tradition dating back to the days of the Alamo and to whom commando operations came as naturally as firing a Garand or BAR rifle.

However, the United States had come to appreciate the value of having a strong SOF capability in the postwar years, when insurrectionist and guerilla movements turned many former Great Power colonies into Third World battlegrounds. This form of warfare came to be known as low-intensity conflict (LIC) to distinguish it from the total global warfare that the United States and her allies had fought since the dawn of the twentieth century.

LIC called for new approaches to engagement of enemy forces, approaches which gave a greater emphasis to the use of special forces troops. The age of the "snake eaters" had arrived.

If warplanners and military strategists are correct, LIC will be the trend for warfare in the twenty-first century, where regional wars and unconventional attack forms will take the place of the global "all-or-nothing" confrontations that marked two world wars and formed the basis of the Cold War.

Special operations forces will undoubtedly play important roles in these wars of tomorrow, as they have done in the Gulf. During Desert Storm, SOF units were equipped for long-range infiltrations of Iraq and occupied Kuwait, for illuminating targets with lasers to beckon to homing precision-guided bombs, for rescue missions, and for other missions that are still classified. The most interesting of these units were special "fusion cells" made up of combined U.S., British, and French commando elements and equipped for extended penetrations of enemy territory on convert missions.

One of the reasons SOF will make major contributions to tomorrow's war is that they are well suited to the "soft-kill" tactical option currently a mainstay of U.S. combat doctrine. This is because SOF elements are well equipped to use nonlethal disabling technologies (NDT) as weapons.

As the name implies, NDT weapons disable but do not kill either humans or mechanical combat equipment. Their sole function is to render both unable to function in combat and so take them out of the fight. NDT weapons are also useful in LIC engagements because they generate low collateral damage—damage to noncombatants and civilian infrastructure. Stray bullets or shell fragments can kill civilians as well as combatants, but NDT weapons are largely benign.

There are currently under development many NDT weapons using a variety of principles. Among them are low-energy laser weapons (LEL), isotropic radiators, infrasound, polymer arresting agents, and visual stimulus and illusion (VSI). Each of these NDTs either restrains or tampers with the human nervous system, incapacitating human combatants without the use of toxic chemical or biological agents, and their effects are easily reversible.

Against war machinery, an array of NDTs, including the aforementioned LEL, liquid metal embrittlement (LME), supercaustics (C+), anti-traction technology (ATT), and combination alteration technology (CAT), are all under development. These NDTs have a disabling effect on mechanical or electromechanical systems, either by damaging them structurally, crippling their ability to move, or incapacitating their combat sensor systems, blinding and deafening them.

The net effect of all these advancements is that on tomorrow's battlefield, casualty rates may drop, and the word "kill" may need to be redefined to include the effects of lethal and nonlethal weaponry both. The concept of warfare will also need redefinition, since it will likely come to resemble the police and prison systems on a transnational level more than conventional military operations of the past.

Rogue countries will be found guilty by a world tribunal, then given the chance to reform or be punished. If they don't reform, they will be rendered powerless by high-technology military attack on many levels. Once powerless, they will be subjected to economic and political sanctions and placed on lengthy probation.

If Iraq and Libya—the latter now trying to become a tourist trap—are any examples, this should be more than enough to keep most rogue states in line.

FOUR

• • • • • • • • • •

Synthetic Battlespace

BEFORE TOMORROW'S SOLDIER hits the beaches of the twenty-first century's hot zones, he'll have to train for the job. The extent and duration of the training will depend on the mission, the military unit, the caliber and capabilities of the troops that undergo it, and other factors.

One thing, though, is dead certain: whatever the mission and whoever the combatant, tomorrow's soldier will receive the best and most thorough training in the history of warfare.

Technological advances in computing and electronics, including virtual reality (sometimes called "virtual environments" by the military), will ensure that the warfighters of the next century are as good to go as it is possible to make them.

Because combat forces are most effective when they can move rapidly and strike with proactive speed, and because modern weapons systems are constantly increasing in complexity, simulator training has become a vital necessity. Even so-called "launch and leave" weapons systems require far more expertise in deployment than the name might suggest, and training with operational systems is neither cost effective nor safe.

Tomorrow's soldier will train in digitally networked modules that will create a lifelike synthetic battlespace in which incapacitation, wounding, and medical evacuation all can be simulated to prepare him for the real thing.

Though research and development of modern simulator systems has represented billions of dollars in outlay, these systems are relatively cheap compared with the expense, danger to personnel, and crudity of technique in training forces with operational platforms in real-world combat situations.

The highest-end stand-alone combat simulators have been developed to train aircraft pilots and mechanized warfare crews. Maritime forces, such as the crews of nuclear ballistic submarines and Aegis battle cruisers, can far more easily patch directly into the sophisticated computer infrastructure of their combat platforms which have built-in facilities for conducting simulated field exercises. But this may change as combat systems grow in sophistication and interservice jointness takes hold.

The Past as Prologue

In the months preceding the landing of American infantry soldiers in Southern Italy in 1943, U.S. troops in boot camp trained with dummy rifles crudely made from planks of wood and mock artillery field pieces thrown together from wood and scrap metal.

To duplicate tanks, half-tracks, and other mechanized armor, jeeps, tractors and other civilian vehicles were crudely veneered with wood and canvas. The result of these training practices was a U.S. invasion force that was ''mean but green'' and that hit the beachheads of Anzio and Salerno largely without having fired a shot in anger.

When tomorrow's soldier hits the ground running to fight the next regional outbreak of war, he will not be similarly unprepared. The Department of Defense has learned many lessons and come a long way since the days when the U.S. Chiefs played junior partners to the British General Staff.

At every level of the combined arms battlefield of the near future, military personnel will have been precision trained for the specific combat environment they will serve in through the use of high-technology simulation and modeling systems relying upon virtual reality interfaces to create synthetic battlespace.

All three branches of the U.S. armed forces currently have programs underway to develop training applications along these lines. Not only will combat personnel train in synthetic battlespace, but computer simulation will also play a part in the combat theater itself, supplementing, enhancing, and in some cases replacing information that in the past would have flowed in through the five human senses. Finally, digital simulation will also be used as a weapon of warfare in the real world.

Virtual Reality for the Military

Virtual reality (VR) has had only middling success commercially, which is not surprising. For all its avant-garde cachet, VR is a technology that was developed by the military as a command and control interface between the soldier-in-the-loop and increasingly sophisticated weapons systems.

The driving force behind military VR research and development is that replacing switches, buttons, throttles, and other controls with virtual simulation makes those complex systems easier to handle by a human nervous system that can process only so much information flow, especially during the stress of combat.

In the wake of the successes gained in the Gulf War employing high-tech weaponry, the tradeoff now more than ever is in destructive power versus accuracy of delivery of the ordnance.

The Pentagon realizes that the U.S. Congress will mandate funds for systems promising maximum lethality on a point target with little collateral damage. VR, then, is like a genie let out of a bottle: make a wish and any system you can name can go from brilliant to Einstein on the fast track of VR-aided accuracy.

A brief discussion of what VR is and how it differs from conventional simulator technologies might put the subject in better perspective. After all, aren't all simulator/trainer systems by definition virtual reality environments? The answer, in the spirit of the Zen koan, is both yes and no.

Virtual reality can be defined as a simulation technology enabling users to immerse themselves to varying degrees in an artificial environment and to interact with objects in that environment.

By means of the technology, abstract data and non-pictorials such as temperature gradients, time flow, radar envelopes, orbital decay rates, depth, and sound can all be manipulated as objects by users. Therefore VR offers cognitive enhancement and skill-mapping via direct involvement through an interaction with objects and environments and a direct participation in procedural flow.

The ability of the human nervous system to process image-based data has been determined to be far greater than its capacity to absorb print and information presented in traditional linear structures. It has been estimated, for example, that the human brain can accurately retain only about 10 percent of what has been read and about 20 percent of what has been heard, but 90 percent and up in active involvement learning. The advantage of VR is that information absorption, management, and interchange are nearly instantaneous and tend to occur on a deeper perceptual level when data is manipulated in objectified nonlinear form.

It should be clear from the above that the key features of VR to keep in mind are *immersion* and *interaction*. It is the degree to which these two components come into play that primarily sets VR apart from conventional simulator technology.

Whereas in a conventional flight simulator, for example, a trainee might utilize a physical joystick, HOTAS (hands-on throttle and stick), or sidestick controllers, an immersive VR simulation system would generate virtual controllers. These would feel rock-solid to the pilot, who would interact with the controls in a normal manner, although in fact they were electronic or "virtual" objects.

However, such a VR interface, because it is in essence nothing more than a construct of ones and zeroes pumped through a computer processor, can be altered with a keystroke (or a virtual pointing device, such as a spaceball) as well as involve sensory processes beyond the normal range.

Threat balloons, radar coverage envelopes from

surface-to-air missile (SAM) sites, location of fighters flying combat air patrol (CAP), ballistic tracks of incoming or outbound missiles, and the like can all be as "real" to the pilot as landmarks on the ground, other planes, or, for that matter, the pilot's own virtualized controls.

Additionally, the pilot can interact with the system in different ways, such as by manipulating the abovementioned spaceballs to click on missile icons or by using eye movements or even thoughts (brain patterns can be scanned using evoked response potential [ERP] technology) in order to achieve the same end.

Since directional orientation in cyberspace is artificial, and the computer is the arbiter of what the pilot sees, hears, and feels, strike baskets can become objects in the same sense as the ordnance being put on target.

In the realm of military simulator applications of VR, the statement often made by its civilian adherents that VR is not "just goggles and gloves" is reversed completely.

While more familiar procedural simulator systems (the huge "Imperial Walker" trainer modules) attempt to isolate the trainee in a physical environment that seeks at least to approximate that which is to be encountered under actual combat conditions, VR systems create an internalized environment in digitized space.

In fact, one of the U.S. Department of Defense's goals in head-mounted display (HMD) development as far back as 1979 had been to reduce both the cost and physical size of military simulators; the capability to project imagery directly onto the retina could eliminate large screens and bulky projection systems.

Military VR Nuts and Bolts

On the hardware end of the picture, a typical military VR configuration includes three main components. The first is a head-mounted display, or HMD.

An HMD consists of three basic elements: image generators commonly utilizing small displays such as charge-coupled devices (CCD), cathode ray tubes (CRT) or liquid crystal displays (LCD), and currently supporting CGA and VGA graphics at the highest end. Because of resolution problems with LCDs, military HMDs predominantly employ CRTs mounted near the ears and mirrors to reflect the image into the viewer's eyes.

This arrangement has the added advantage of permitting dual usage as head-up displays if the optics are semi-reflective but has the disadvantage of placing high voltages extremely close to the wearer's head.

Ultrasonic, mechanical, optical, or inertial head position trackers map viewing perspective into the HMD's computer processor memory. These utilize a six-degree-of-freedom (6DOF) principle to measure and define position and orientation by means of calculations of three translational and three rotational values along x, y, and z axes comparable to the roll, yaw, and pitch used in flight terminology. A rear counterweight to provide stability completes the standard HMD assembly.

Many of these devices are available commercially, although it's not unfair to state that almost every U.S. manufacturer has at one time had a defense connection—indeed, most would still be servicing DARPA contracts if economic pressures had not forced them out into the private sector. Polhemus Incorporated, for example, one of the earliest manufacturers of a type of head-motion tracker commonly found in HMDs, has recently been awarded a contract to manufacture head-mounted sight trackers for adoption in HUDs used in the Commanche helicopter fire control and navigation system.

As for the two other main components of a VR system, force feedback gloves (FFG), sometimes called "wired gloves," and less commonly an environmentally insulated "cybersuit" such as the VPL Datasuit, are used in tandem with the HMD when virtual object manipulation and/or full-body immersion are called for.

The FFGs—most based on the VPL Dataglove model—enable users to manipulate objects in the virtual world. Some designs involve gloves that are air-filled to simulate weight and other tactile features of virtual objects; these are tactile feedback gloves.

Fiber optic or electromechanical sensors in both gloves and suit use 6DOF protocols to translate motor responses into digital data processed by the computer processor in order to facilitate interaction with the virtual environment.

Objects can be manipulated as well as physically altered, overlayed upon, and/or combined with non-pictorials to highlight specific areas of interest. Force-feedback handgrips of various designs and configurations, as well as wands, mice, and force-balls, are also used in this capacity.

These systems have obvious important implications for real-world as well as training applications, which in medicine are already paying off in improved surgical procedures. If a surgeon using VR can pinpoint a tiny cancerous cell and zap it with a laser, much the same can be true of an F-117 pilot about to put ordnance on target, a submarine weapon system officer

(WSO, pronounced "Wizzo") about to launch a torpedo, or an AWACS or JSTARS radar systems officer
(RSO, pronounced "Rizzo") detecting a launch signature in the air or mechanized armor movement on
the ground.

The ultimate weapon of the future may not be the
one that will blow up the world but one that will
merely blow up Saddam, Ghadhafi, Milosevic, or
whoever happens to be the maximum bad actor of the
moment, preferably without staining the carpet. One
day, possibly not too far off, the famous philosophical
problem concerning the push of a button resulting in
a fatality 3,000 miles away may not only be a testable
proposition—utilizing VR's capabilities, it might
have become tactical doctrine.

Where Synthetic Battlespace Gets Real

If you're beginning to get the impression that this
discussion has veered somewhat from simulators to
real-world applications, then you are absolutely correct. As should be evident from the above, VR
technologies can share the properties of both simulation and control systems, and they can also be used
offensively.

In fact, the development of military VR systems
was never undertaken purely as a better means to
teach pilots to fly planes; it was meant to be a command and control (C2) interface to assist them in flying them. It was never intended to effect economies
solely by training Tomahawk crews using simulated
ordnance; it was meant to enable the "Wizzos" to
better guide their rounds to the intended targets.

VR had not been intended to make sonar operators

more astute at detecting submerged or surface contacts, but to automate military personnel and the interactions between military personnel and the systems which they control in much the same way that weapons systems have been automated over the last twenty-five to thirty years.

Actually, there does seem to be an approximate thirty-year development cycle in high-end U.S. military systems. Witness stealth technology, for example. The prototypical Stealth bomber, the YB-49 "Flying Wing," required some three decades before the nascent technology matured into the B-2 Spirit. If such is the case, then VR-based systems, first introduced circa 1963, are now poised for incorporation into mission-critical systems.

In fact, among the U.S. Defense Department's chief research and development goals for the decade has been the development of synthetic digital environments in which computer modeling and simulation would serve roles from rapid prototyping of advanced weapons systems down to the molecular level, and conversely, tracking the course of a threat-rich combat environment down to the ballistic tracks of individual warheads.

Utilizing massively parallel processors (MPPs) that can at present perform some one billion operations per second (OPS) and which are projected to show a tenfold increase in processing speed within the next three to five years, digital modelling has been applied by Defense's Supercomputing Alliance to clarifying images generated by synthetic aperture radar (SAR) in order to pinpoint and identify weapons platforms otherwise lost in electronic clutter, and to do so at high speeds. The best SAR images are classified, but these radar "snapshots" are said to be so good, they're indistinguishable from high-quality color photography.

Modeling and simulation technologies utilizing MPPs also have applications in wargaming simulations to replicate theater combat situations accurately, increasing realism and anticipating otherwise unforeseen tactical and strategic developments with a high degree of accuracy. In the realm of networked environments, the Pentagon's Joint Warfare Center utilizes interactive hardware and software in a digital video branch exchange (DVBX) that can involve thousands of participants over long-haul computer nets spanning the globe.

On a related front, strategic/tactical mission planning and support systems utilizing computer simulation technologies, having proved their worth in the Gulf, are now an essential component of combined arms battlefield doctrine.

The USAF's mission support system (AFMSS) and the USN's special operations forces planning and rehearsal system (SOFPARS) are two programs geared toward extending the technological potential offered by computer modeling of synthetic combat environments.

Synthetic Theaters of War (STOW)

The proliferation of VR-based systems into military simulator environments has given rise to the term synthetic theaters of war (STOW). Collectively, STOW encompasses all non–real world combat simulation facilities in use today. In an era of shrinking military expenditures, overall VR budgeting has been conspicuously spared the swing of the congressional ax. In fact STOW research and development is on the increase.

This is because STOW promises to deliver more bang for the training buck. An HMD, suit and gloves linked to a desktop computer which is in turn networked to a mainframe host, for example, can perform many of the functions previously requiring the use of simulators costing millions of dollars. System elements can frequently be purchased as common off-the-shelf (COTS) technology, facilitating the inexpensive stockpiling of spare parts.

VR is also an enhancement of existing technology. Aging or "legacy" systems need not be retired. With VR systems they can be enhanced or upgraded. VR technologies support what the military calls "cross-platform diversity."

Reconfiguration of simulated mission parameters can be achieved with relative ease by reprogramming the computer system to compensate for incompatibilities in systems utilized by different branches of the armed services. By making planes, tanks, helicopters, communications linkages, and military personnel all virtual objects in virtual space, great leeway in the types of battlefield scenarios is possible.

One virtual reality technology, televisual reality (TR), is especially well suited to mission simulation applications. TR is a networking technology enabling noncolocated participants linked on the net to share cyberspace. TR should not be confused with telepresence (TP), which it closely resembles. Unlike TR, telepresence does not necessarily imply interactions among occupants of the virtual world.

Televisual reality enables distributed processing across large, even global, distances. Using the technology, an instructor located thousands of miles from fledgling sonar operators could appear to be in the same simulated crew station aboard a virtual Seawolf submarine, for example.

The U.S. Navy has for many years been the branch of the armed services on the cutting edge of VR research, development, and fielding of prototype systems, such as teleoperation (TO). This trend can be explained by two overriding factors. The first is the USN's need to maintain multilevel and multilayered force structures (e.g., a carrier battlegroup) arguably unique to the service branch. The second is its need to conduct undersea missions.

In both situations, nonpictorials such as radar and sonar envelopes, depth, isothermals, simultaneous location of hostiles and friendlies, and surface and submerged contacts makes a technology that can effectively integrate these tactical elements critical to the success of the mission. Along these lines, the USN's advanced combat direction system (ACDS) program incorporates advanced data fusion technologies to create an integrated tactical simulation of the operations area for battle group commanders.

Televisual reality simulator technology is central to the operation of the Simnet administered by the Defense Advanced Research Projects Agency (DARPA). With joint DARPA-Army development begun in 1982, Simnet is one of the original and still one of the most ambitious military utilizations of simulations technology for training applications.

Upwards of 270 simulators have been installed at locations in the United States and Europe since the program's inception. These distributed Simnet nodes are linked together via both local and wide-area computer networks (LANS and WANS) of up to 100 individual simulators and also via satellite in the form of a long haul network (LHN) in order to facilitate team processing operations in a networked televisual reality environment.

Patched into the televisual reality network, armor, mechanized infantry, helicopters, and fixed-wing aircraft such as fighters and bombers, as well as FAAD (forward anti-aircraft defense) emplacements, can participate in interactive wargaming utilizing highly realistic audiovisuals.

By means of multiple simulator node linkages, crew commanders can train their units against one another or against OPFOR (opposition forces) units in a wide variety of simulated battlefield engagements over a wide assortment of terrain, including a generic battlefield (DARPA claims any terrain on earth can be simulated by the system), with a "combat arena" of some thirty square miles and local terrain patches of some two square miles.

At the heart of the Simnet system is a management, command, and control (MCC) interface consisting of network clusters linked to a mainframe host platform. These linkages include staff officers overseeing operations in a virtual tactical operations center.

Although it was initially limited to relatively low-resolution polygonal picture elements with faithfulness in representation reserved for military hardware and generic icons serving to connote background objects such as trees, buildings, and the like, advances in image processing and in dynamic modeling technologies and in raw CPU processing speed and power have vastly improved the quality and variety of the simulated environment.

By means of dynamic terrain algorithms, for example, craters can be programmed into the environment following an ordnance strike; other realistic details, like tank treads left in the wake of the passage of mechanized armor across the virtual landscape, can also be displayed.

Real-time update rates for individual simulator nodes have been increased as well, resulting in decreased perceptual lag-times and image density in the area of a reported sixty frames per second. While Simnet continues to expand its capabilities, improved features continue to be added.

However this may be, it is the military's aviation arms that in the aftermath of surprisingly successful air operations over Iraq have reaped the budgetary spoils of victory in the Gulf. The U.S. Air Force as well as naval and marine aviation have been allocated major funding for implementation of VR systems in flight training applications.

USAF research and development has led the way. Since the debut of the VCASS (visually coupled airborne systems simulator) at Wright-Patterson AFB in 1982, with its oversized helmet that came to be called the "Darth Vader," the USAF has actively pursued R&D for systems incorporating VR in both flight simulation and actual cockpit C2 functions.

The new man-machine interface offers the possibility of changing the pilots' favored metaphor of "strapping the aircraft to their backs" to something more akin to "grafting it to their skin" where such asensory processes as radar paints and servomechanical linkages become extensions of the pilot's body and sensory apparatus.

Head Mounted Displays (HMD) in Training and Combat

Today, advanced HMDs capable of displaying high-density image resolution and interacting with an advanced graphical display interface, as well as interactive simulation technologies that link participants across networked environments, are being used in the realm of pilot simulation training. A variety of systems improvements, such as massively parallel processing (MPP), have also made possible a far broader spectrum of mission training sets than were previously in the simulation repertoire, including, for example, pure pursuit maneuvers, where pilots track enemy planes through a wide assortment of jinks, turns, dives, etc.

The USAF, utilizing advanced HMD technology developed by CAE-Link, General Electric, and other technology providers, continues to modify and improve its virtual or "supercockpit" command interface pioneered with the VCASS program.

Here graphical icons, such as SAM threat envelopes and onboard weapons stores, as well as text blocks displaying systems status and other tactical information, are overlayed on real-time video of the air combat zone, replacing traditional HUD functions with an integrated command and control interface. (The overlaying of real-time visuals with digitized graphics is a technological subset of virtual reality known as "augmented reality" or AR.)

But the applications of HMD technology go far beyond training. There is, of course, no point in training F-15 and F-16 fighter crews utilizing such a "god screen" or "supercockpit" command interface, since the planes don't support it, nor are they ever likely to. Then why invest in the technology?

The answer is that advanced tactical fighter (ATF) designs such as the F-22 and Joint Strike Fighter (JSF), upgrades to extant stealth aircraft such as F-117 and B-2, and certain still officially classified high-altitude transonic stealth aircraft (AURORA) may incorporate some or most of the virtual cockpit technology. Tomorrow's generation of warplanes will probably have some form of VR control interface to help the pilot fly them, as USAF white papers concerning the JSF strongly indicate.

Interactive simulation applications utilizing virtual reality technologies will continue to play an increasing role in training military personnel on ever more sophisticated and ever more expensive combat platforms, but there is a downside.

Some DOD-sponsored reviews of VR simulation research and development throughout the decade have warned that VR might not be the panacea for flight training that its adherents claim and suggested that for a number of pilot training programs, more conventional flight simulators and/or training in actual aircraft was advisable.

Simulator sickness is a common drawback cited by VR simulation critics. It is caused by the discrepancy between visual motion cues generated by the system and those provided by the human senses. In interactive VR networks these symptoms can be made worse by the added stresses to the human nervous system brought on by lag-time in screen refresh rates, producing such reactions as headache, nausea, and vertigo. So far, no long-term studies on the effects of synthetic environments on the human nervous system have been made.

All this notwithstanding, synthetic environments for training and combat, including military or offensive VR, will likely come to play increasing roles in tomorrow's warfare. Tomorrow's battlespace will be as much in the world as it is in the mind, as much in cyberspace as down in the mud, as much grand illusion as it is gritty reality.

One theater of combat that will be greatly transformed by information technology will be the land battle. It will be changed forever by such developments as the dawn of hyperweapons and the metamorphosis of the infantry line soldier into the supertrooper of the future, the digitized G. I.

We will be meeting both of these in the pages to come.

FIVE

• • • • • • • • •

Land Warfighting 2010

THE LAND BATTLE. Throughout the history of warfare it has been the yardstick by which an army is measured, the scourge by which its mettle is tested. For every sea battle like Midway and every air battle like the Battle of Britain, there have been a dozen Stalingrads, Hues, and Inchons.

Ours is an age obsessed with air power, in which the threat of a Soviet invasion of Europe has receded and in which U.S. troops have not seen prolonged ground combat since America's engagement in Vietnam.

Ten years ago, the military forces comprising NATO were postured toward defending the West against waves of Russian tanks pouring through the Fulda Gap and rolling across the vast plains of Eastern Germany into the heartland of Europe.

It was considered a virtual certainty that the third world war would begin as a land war, even if it might end in nuclear cataclysm. But with the coming of glasnost and perestroika, superpower tensions began to abate and soon the Soviet Union itself was a thing of the past.

More recently, in the most unprecedented buildup

since World War Two, some 350,000 U.S. troops joined Coalition allies in training for a planned assault on Iraq and occupied Kuwait.

They would be facing the world's fourth largest army and facing it on its home turf. They would have to surmount formidable barriers, including minefields, barbed wire emplacements, and pits filled with burning oil across which dug-in Iraqis could pour lethal machine-gun fire.

An air assault phase of the war, Desert Wind, would precede the land invasion. The planes would soften up enemy defenses, relentlessly pounding away at command, control, and communications centers and at personnel bunkers before the ground phase of operations commenced. The strike aircraft would fly as many sorties as it took, rolling over Baghdad in continuous waves and dropping their bomb loads. Indeed, U.S. pilots dropped over 7,400 tons of precision munitions during the war. The F-117 Stealth fighter alone accounted for some 2000 tons of ordnance delivered during approximately 6,900 hours of combat airtime.

Yet at some point Coalition ground forces would have to move in, and when they did, it was considered a given that they would encounter stiff enemy resistance. There would be casualties. Probably some units would be hit hard. The line grunts who awaited the go order had not forgotten the old army maxim that "infantry die."

In the end this eventuality did not develop as anticipated. Instead, the air assault decimated the enemy defense networks, reducing enemy military infrastructure to twisted wreckage with surgical precision and pounding the daylights out of dug-in troops with nonstop punishment.

When the campaign's ground assault phase commenced, it was a walk in the sun that nobody expected. Even tank units of the elite Iraqi Republican Guard proved no match for well-drilled Coalition tankers equipped with M1A1 Abrams and British Challenger tanks.

What transpired in the Gulf precipitated a reevaluation of air power's capacity to unilaterally dismantle the enemy's capacity to wage war. For the first time in history the basic premise that a military force cannot win a war without waging a land battle proved an outdated concept. It was a revolutionary turning point in the history of warfare, perhaps as radical as the replacement of troops by artillery to breach the defended walls of Renaissance cities had been hundreds of years before.

Tomorrow's war may see this new vision overturned in favor of the old. Tomorrow's enemy will have also studied the Desert Storm scenario and decided that had the Iraqis hunkered down and fought like men, things might have turned out somewhat differently.

Having learned this lesson, tomorrow's enemy might decide to avail itself of a number of high technology defensive measures, from newer and smarter portable rocket launchers to faster, stealthier tanks. In tomorrow's MRC, this newer, tougher, and better-equipped enemy might just give our forces a run for their money, despite the deadly steel rain we can pour down on it from the air before we strike overland.

Tomorrow's army will need to train for a worst-case scenario. Nothing less makes sense. At the same time, peacekeeping missions like those in Somalia or Haiti will require the presence of ground forces who are ready for any and every eventuality that might arise. Such operations have had a knack for turning ugly very quickly and degenerating into shooting matches in urban quagmires.

Finally, ground forces will continue to play greater and ever more critical roles in ballistic theater missile attack and defense scenarios. As incredibly expensive airborne assets such as Joint Stars and stealth aircraft continue to be fielded, ground defense installations will be needed in ever greater numbers to protect these assets from enemy strikes aimed at decapitation prior to takeoff.

Theater missile installations will also give tomorrow's ground combat forces a critical over-the-horizon capability of projecting offensive military force as well.

Ground-based installations will send out UAVs on reconnaissance missions, often at long range and high altitudes. These forces will use the data gathered from such drone flights to plan and execute missile strikes against distant targets. These may be stationary targets, such as enemy C4I nodes, or mobile targets, such as tank formations or even seaborne task forces.

The line grunt of tomorrow—the dismounted infantryman, in military usage—will change too. In an era of an all-volunteer army, and one where friendly body counts and high casualty rates are not tolerated by the electorate on the home front, the infantry soldier will be "upgraded." If he will continue to be a highly mobile, MRE-fueled, good-to-go weapons platform, he will be a far more survivable one, a supertrooper encased in a high-technology cocoon.

Today's line dog is already better protected than ever before by composite body armor and his combat punch is amplified by superior arms. Tomorrow he will also be digitized and Internetted, transmitting and receiving high-bandwidth visual and audial data to and from distant command and control centers, orbiting military satellites, and members of his own platoon, company, and battalion.

He will field computer-targeted man-portable weapons systems that will increase his firepower by a factor many times that of his twentieth-century predecessors. He will interact with mechanized armor, warplanes, and vessels at sea to call in air strikes and artillery support when the going gets tight.

He will bleed and die as well, but in fewer numbers than ever before. And when he does suffer battlefield casualties, advanced combat medicine and evacuation technologies will be available to help reduce the impact of those casualties so he can be sent home for treatment or back into the fight.

Tomorrow's Land Warfighter: The Digitized G.I.

The line grunt has been the basic low-tech weapons platform and forward sensor system of all military units since the dawn of time. He's the universal soldier and he's been around for centuries.

Correspondingly, there have always been plans to upgrade, enhance, advance, and otherwise improve the combat lethality, mobility, and survivability of this organic-fueled, bipedal-locomoted, and constantly bitching biological combat element.

That most of these plans haven't amounted to a hill of beans can be judged by the fact that today's dismounted infantryman is pretty much the same general-issue line dog as his Vietnam-era counterpart, or, for that matter, as his BAR and Garand-cradling forefathers who slogged their way from the Normandy beachheads to the heart of Berlin.

All the same, high-tech initiatives toward "aerospacing" or "Buck-Rogering" the line grunt along lines similar to the technological advances that have gone into fighter planes continue to be developed.

Advanced warfighting experiments or AWEs in support of the U.S. Army's Force XXI initiative have fielded ground forces linked by C4IFTW—C4I for the warfighter—technologies in which advanced sensor-to-shooter networks afforded a shared situational awareness and real-time battlespace synchronization with troops in the field, commanders in the rear, overhead assets, such as helicopters and surveillance aircraft, and mechanized hardware, such as tanks, in theater.

There is clear consensus at the Pentagon that the infantry soldier will play an enhanced role in future combat. The concepts of jointness and of the digitized battlefield both visualize twenty-first century warfare as an integrated whole, so the grunt on the ground, planes in the air, and ships at sea will all have to be in the same loop, part of a grand "system of systems."

Also, with off-the-shelf dual-use computer components available that can be inserted as they are into military systems or cheaply and quickly redesigned to work with existing hardware, manufacturing small, smart, and survivable computerized systems that can be carried around in battle by the individual soldier is more doable than ever before. These systems could detect the presence of hidden enemy forces, upload video data to satellites, target the infantryman's weapons, and inject him with counter-NBC agents, stimulants, and nutrients.

Prototypes of the Buck-Rogered G.I. have been in existence for years, and there are current programs in the United States, United Kingdom, and France to cocoon the dismounted infantry fighter in a hard-to-kill supersuit and link his weapons and communications to networked computers.

The Pentagon's Land Warrior 2000, the British MOD's Infantryman 2000, and the French ECAD Advanced Combat Soldier System (*Equipment du Combattant Debarque*) all share similar basic goals—and to one extent or another, incorporate common transnationally developed component systems.

This Buck-Rogered, aerospaced grunt—let's call him the "supertrooper"—would be enveloped by a multilevel offensive/defensive and communications system made up of several component parts or layers.

The first layer would be the supersuit itself.

This supersuit's exterior would be made of lightweight ballistic armored material, with extra ballistic protection built into the lower leg area to protect tomorrow's soldier against land mines.

The suit would be stealthy as well as kill-resistant. Various means to reduce its infrared signature to hostile sensors would be used, including a built-in cooling system in which a coolant substance flowed through the outer skin via a network of plastic capillaries.

The suit could also be chameleonic, sprayed with biodegradable colors that could change its shade and patterning. It might also use a more advanced system that wove liquid crystal display technology into the fabric to effect pattern and color change. In forest, the suit would adapt the color of the underbrush; in desert, the color of sand, rocks, and arid vegetation. It would adapt to match the environment.

The suit's interior would have its own microclimate—cool in hot weather conditions, heated in cold, and sealed against invasion by air-delivered chemical or biological agents.

The supertrooper's supply of breathable air would be filtered and purified for rebreathing and his solid and liquid wastes collected and chemically sanitized for recycling and disposal during downtime. Biosensors would monitor the supertrooper's physical condition, injecting him with drugs, nutrients, or medicines if he were tired, hungry, or wounded.

Other sensor systems, linked to voice and brainwaves, would make up another layer of the system. These sensors would both provide real-time tactical data and enable voice- and thought-activated control of onboard systems.

Forward-looking infrared (FLIR), laser, low-light video, global positioning system (GPS) telemetry and audio data would all be routed to an onboard central processor and fused into a comprehensive tactical picture visible on a head-mounted display (HMD) inside the helmet and audialized via three-dimensional stereo sound.

The HMD would display a mixture of real-time video of the battlespace and an overlay of graphic and alphanumerics representing threat envelopes, targeting aimpoints, and digital map information. For example, a full-screen, live-video image of a cleared area surrounded by trees would be overlayed with a green wire-grid pattern showing the safest route through the clearing.

Red symbols spaced along this route would indicate the rated probability of ambush or of mines or other countermeasures being engaged. In ghostly white, the concealed presence of any radiating object—the moon behind treetops or a stealthy helicopter—would be part of the view.

Alternatively, a pop-up window could contain an aerial view of the same clearing from orbital space, courtesy of a spy satellite, or a closer-in look at some questionable target from the cameras in a micro-UAV that the supertrooper hand-launched like a toy glider moments before.

Another window might contain the face and voice of the supertrooper's mission commander who is situated in a mobile base miles away, with voiceprint and retinal pattern identification to enable him to make sure it was really the commander and not a virtual reality simulation intended to bring him to harm.

The HMD and the portable computer that feeds it data would also be linked to the supersuit's third layer, its main weapons system. This would be a multipurpose weapon; a combat rifle, multiple grenade launcher and general purpose machine gun all rolled into one.

It would be targeted and, under certain conditions, automatically fired by the suit's onboard fire control system. Some plans call for a backpack missile launcher that could fire both ground-to-air rounds for killing helicopters and top-attack missiles—those which deploy downward from the air after launch—to stop tanks and other mechanized armor.

As farfetched as all of the above might sound, a great deal of research and development is being poured into prototype systems based on this model. At some point, much of the futuristic supertrooper will become a reality. When that will happen is anybody's guess right now, but by the year 2010—about eleven years from now—more of this technology will be on the battlefield than present-day critics believe.

After all, in the last thirteen years the U.S. infantryman has seen a return to the body armor of medieval knights in the form of ballistic protection; he has donned the so-called "Fritz" helmet, which is an exact replica of the German Wehrmacht helmet of World War Two, albeit in Kevlar instead of steel; he has used computerized communications and positioning systems in the form of SINCGARS and GPS; and his firepower has increased at least 60 percent, with the adoption of man-portable missile systems. The land warrior of the twenty-first century will likely resemble the G. I. of World War Two or Vietnam as little as these soldiers resembled the infantrymen of the late nineteenth century.

Grunt 2010

Just how much tomorrow's soldier will change in order to mesh with Force XXI's digitized battlespace vision is an open question. Technological questions aside, there is ongoing debate and legitimate concern about high-technology enablers turning G.I. Joe or G.I. Jane into G.I. Nerd and G.I. Woos.

Historically, the prime weapon of the infantry has been raw courage, staying power and will to win. This "right stuff" of the Audie Murphy variety, more than anything else, has made the difference between winning and losing, fulfilling or fumbling the mission—indeed, between living and dying itself.

There is little doubt that the Buck-Rogering of the infantry soldier will, in fact, have this effect to greater or lesser degree, if for no other reason than that using the new digital combat systems effectively will require more technical skill than brute strength.

But would this turn of events reduce the warrior to the push-buttoneer?

Yes and no.

Let's look at the historical record. At the Battle of Marathon in 490 B.C., the Athenian Greeks faced off against the Persians in one of the decisive battles of history. This type of battle usually was over in under an hour, but it was nevertheless the most brutal form of warfare ever practiced—combatants were compressed into a stabbing, slicing, hammering mass with no escape until one side either was slaughtered or fled.

By contrast, the post–World War Two Salvo study documented that during the war, troops on both sides of the conflict broke and ran when their rifles jammed or their ammo supply was depleted and bayonets or bare knuckles would need to be used. Bombs and bullets at long range accounted for most combat fatalities during the "Real One." During the Gulf War, ground troops did even less fighting, herding an already broken and defeated enemy who had been saturation-bombed to near senselessness by concerted air strikes.

This is not to disparage the infantry soldier in any way, merely to point out that as the nature of warfare changes, so inevitably does the nature of the combatants who fight it.

The record shows that as time has passed, close-in engagements have become long-range killing contests using more lethal and more accurate weaponry which have in turn required more technical knowledge and skill to deploy. If the trend continues, and there is every reason to believe that it will, the supertrooper of the next century will eventually come into his own.

Admittedly, it's not likely that the supertrooper will stalk tomorrow's near-future battlefield. In our warfare scenarios for the years 2010 to 2020 we'll have to consider some more moderate developments that will enhance the warfighting capabilities of the infantryman.

Advanced joint high-bandwidth networked tactical communications systems will certainly be part of these enhancements. Some systems—such as SINC-GARS, a compact communications unit displaying visual and voice data, and GPS, a global positioning system using downlinked data from a network of orbiting satellites—are already in common use on the battlefield.

Portable computer-targeted weapon systems of various kinds that extend the firepower and range of the infantry soldier by many orders of magnitude, such as Dragon, Strix, and Stinger, are also in use, with newer systems on the way.

Fast, light armored combat vehicles of various types will move tomorrow's land warrior to and from the combat zone and provide platforms for weapons and surveillance systems. Standoff robotic systems, such as small, hand-launchable UAVs, will give him wide-ranging reconnaissance capabilities.

Tomorrow's infantry soldier will have better communications than his Gulf War forerunners. These will be global-mobile (GLO-MO), meaning they will be integrated at all levels of the tactical and strategic battlespace. GLO-MO communications devices will be smart and portable, able to fuse data from several sources into a comprehensive whole.

One such device will be the tactical battle assistant, or TBA. These will be hand-held units capable of displaying real-time visuals and graphics at high resolution. Tomorrow's soldier will use his TBA to communicate with his rear, to show him his position on a map, to update him on the enemy's position, to calculate fire solutions, to call in close air support, and to perform a variety of other tasks.

If he breaks off from the platoon individually or as a member of a special action squad—the mission might be to target mobile Scud launchers and call in an air strike—he will use small hand-launched unmanned aerial vehicles to perform reconnaissance and transmit fire control data to a distant artillery unit or an in-flight attack helicopter force using the TBA or other GLO-MO devices.

Tomorrow's infantry weaponry will also change. The assault rifle will remain, but it will be more accurate and have a higher rate of fire than past versions. Many programs to develop successors to the M16 and other conventional rifles have been undertaken. The CAWS (close assault weapons system) program developed by Heckler, Koch, and Olin has produced the G-11 caseless assault rifle, for example.

That the G-11 is a radically different weapon is obvious at first glance. Its short "bullpup" design blends barrel, stock, trigger housing, and other components into a compact whole of molded high impact plastic. Its ammunition clips look more like Kit-Kat bars than standard magazines—they are blocks segmented off into plastic sticks.

Each of these segments is a caseless round, so named because the bullets use no cartridge. The 4.73-by-33 millimeter (mm) ammunition is more accurate and has less recoil than the current standard NATO caliber 5.56 and 7.62 mm rounds presently in use. The G-11 has already entered service with the German Army and elite units of the U.S. Armed Forces.

The soldier's personal sidearm—yesterday the Colt M1911 and today the Beretta M92—may also be superseded by something like the P-90 personal defense weapon (PDW) developed by Fabrique Nationale and also currently in service with special units worldwide. The P-90 is an unclassifiable weapon, part pistol, part submachine gun, and at the same time something of an assault rifle, too.

The P-90's radical design also marks it as something special. Its black molded plastic exterior resembles two doughnuts welded to a hatchet and sprouting a short ugly muzzle. Unlike other weapons, the P-90's clip of fifty 5.7-by-28 mm rounds loads horizontally and the spent shell casings are ejected from a port at the bottom of the weapon.

The nonstandard ammunition used by the P-90 is said to have three times the neutralizing power of a standard NATO 9 mm round, and the downward ejection principle used by the gun is said to make it nearly recoilless, all the better to hit something for tomorrow's soldier. In fact, the 5.7 mm round should very soon be adopted as the NATO standard, replacing the 9-by-19 mm "parabellum" round for military use, a round originally developed in Germany during World War One.

For bigger game, such as main battle tanks, enemy aircraft, and rotorcraft, and even large troop contingents, tomorrow's line dog will have at his disposal a number of heavy firepower producers that will give him an over-the-horizon—as opposed to line-of-sight—kill capability.

This means that these advanced weapons will use satellite and other tactical data to target distant armor, load the fire solutions into the man-portable system, and enable the infantry soldier to fire and forget from the comparative safety of a concealed position—"comparative" because the enemy will have its overhead assets, too, and these will be looking for him.

Today's premier tank killer, the TOW missile, is a wire-guided missile that needs a direct line of sight link between shooter and target for at least three seconds.

While the TOW round is in flight, the shooter must keep it fixed in his target reticle while signals transmitted along the wire paying out from its rear keep it on a steady course. This not only makes TOW a less reliable weapon for use against agile targets maneuvering at high speeds, it also exposes the shooter to counterfire for approximately three seconds, and in combat that can be a lifetime.

Finally, TOW is another weapon system originally designed for World War Three battles fought against Soviet tanks using Soviet-style tactics in a European-style battlefield. While the arid desert terrain on which the land phase of Desert Storm was carried out was very close to the conditions for which TOW was first designed, these may not be in existence in tomorrow's MRC.

Three systems that give the shooter launch-and-leave capability are almost sure to be upgraded variants of the SADARM, Strix, and BAT systems. These systems are known as top-attack systems because they launch their missiles into a vertical trajectory.

Once the round reaches the top of its trajectory envelope, sensors onboard the warhead locate the target, commonly a main battle tank or other heavy armor. The sensors are smart enough to distinguish intact targets from burning or partially destroyed ones and to pick out the real McCoys from the decoys.

Once the sensors acquire, the round then goes into its terminal phase. Small rocket engines in the warhead activate to send the round hurtling downward, correcting its course in case the target takes evasive action. At the point of terminal engagement, just before detonation, an armor-busting primary charge goes off to breach the hull of the target vehicle, followed microseconds later by the explosion of the main shaped charge that injects a jet of molten metal into the vehicle, blowing large holes in it and killing any living thing inside.

Chariots of the Grunts: Tomorrow's Ticket to Ride

Tomorrow's infantry will have a faster ride than ever before to where the fighting is heavy. Strategic and tactical air- and sealift will carry these forces to within debarkation range of the battlezone, and drop them into the center of the combat theater.

Heavy transport ships like the mammoth Hubble-class ships capable of carrying two thousand troops along with supplies for thirty days will accompany carrier battlegroups steaming in from the sea. Heavy-lift aircraft such as the C-5A/B Galaxy, the C-130, and the new C-17 transport will fly the troops inland.

Air-cushion LCACs will transit from the littoral to the land-based forward deployment areas. The advanced V-22 Osprey tilt-rotor hybrid aircraft will shuttle combat personnel rapidly from the beachhead to deployment zones on the strike perimeter.

Once in-theater, MICVs—fast, armored carriers—dubbed by some "battlefield taxis," will shuttle the now "mounted" infantry (sometimes called "cavalry") to points along the perimeter where they will set up firebases, scout out terrain, set up ambush- or hide-sites, and engage in other designated tasks in support of the overall mission.

The MICV (mechanized infantry command vehicle) is the modern infantryman's equivalent of a team of combat workhorses. The U.S. Army's M2/M3 Bradley armored carrier is the world's foremost example of the hardware category. The Bradley is replacing the older M113 vehicle, of which some 73,000 units have been built. It has been the basis for most MICV designs worldwide for some two decades.

The Bradley has a lot going for it, including high mobility, provided by its eight-cylinder diesel engine, high-tensile strength laminate armor covered by reactive armor plates, and the heavy firepower afforded by its TOW antitank, LAW rocket launchers, and 25 mm cannon. The Bradley also features a 7.62 mm coaxial machinegun and 40 mm grenade launchers for secondary armament. Fully amphibious, the Bradley carries up to seven troops with full battle gear in its cavalry configuration and can travel at speeds over 41 mph on open roads.

All the same, the Bradley is yesterday's combat vehicle, and tomorrow's soldier will probably view it as today's soldier views the older M113—as a military system on its way out.

This is because despite the Bradley's many acknowledged strengths, it is essentially a pre-RMA piece of military hardware, intended to perform in a World War Three combat environment in which maneuver warfare on the open plains of central Europe would be the theater in which it would operate.

Tomorrow's major regional conflict will probably not resemble this defunct combat scenario. Twenty-first-century wars will require a different type of vehicle, one that is more agile, stealthier, lighter, and smaller than the Bradley, yet one that can carry a full twelve-man platoon and offer serious fire support to troops on the ground in a broad range of tactical situations.

For want of a more precise term, this vehicle is currently referred to as the FIFV, the future infantry fighting vehicle.

Unlike the Bradley, whose target silhouette is close to that of the M-1 main battle tank, the FIFV will ride on a low-slung angular chassis that is composed of lightweight composite armor. Its hull will have radar absorbant properties, giving the vehicle a tiny radar cross section.

The FIFV's exhaust and engine system will also be especially designed to reduce its visibility to infrared scanners. Unlike the Bradley or M113, the vehicle will not have a turret. Its armament will be conformally recessed into the hull and raised when the enemy is engaged, somewhat like that of advanced tactical fighter planes.

Nor will it need a dedicated driver or require additional space for control mechanisms. The FIFV's advanced vetronics system will control all aspects of the futuristic war wagon by a virtual environment system using wired gloves and a head-mounted display.

The FIFV will be harder to see and harder to hit by the smarter, more agile antiarmor missile systems that will be encountered on tomorrow's battlefield. Since it will be smaller too, there will be two or three targets that enemy rounds will need to acquire instead of one slower, lumbering target.

Tanks, Artillery, and
Other Mechanized Assets

Most of what's been said about the Bradley also holds true for another contemporary leader in battlefield hardware, the M1A2 Abrams main battle tank (MBT). The Abrams proved itself in the Gulf War against the best armor fielded by an army equipped and trained by the former Soviets.

Iraq's fleet of Russian tanks, capped by the top-of-the-line T-72, proved to be no match for U.S. M1 and M60 series tanks, which outshot, outmaneuvered, and outlasted them in battle. In fact, in the aftermath of the tank battles of Desert Storm, the former Soviets, watching from the sidelines, were faced with the unpleasant realization that their best tanks were outclassed on every level by the U.S. tanks.

Unfortunately for the Abrams, this is one reason why it is doomed to obsolescence in any conflict with a well-armed regional enemy. The appalling performance of the Russian MBTs has been a lesson learned the hard way, but the former Soviets immediately began rethinking their tank design strategy in general and about how to deal with American armor in particular.

The result of this thorough rethinking has been new tanks that better match the capabilities of U.S. counterparts and are equipped with better gunnery than before, gunnery that more closely matches the capability of the M1s to fire high-technology rounds including the APFSDS kinetic energy (KE) round.

These are "saboted" rounds, which have a core of some very dense metal, such as depleted uranium, sandwiched between layers of propellent—"sabots"—

which impel them from the gun barrel at extremely high speeds.

The rounds themselves use no explosive to cause destruction. They don't need explosive to kill armor.

When kinetic energy rounds strike the hull of a tank or other armored vehicle, the energy they release instantly liquifies the hull at the point of contact, injecting a stream of molten metal inside, maiming flesh, and blowing up fuel and ammo stores.

But today's armor is getting better at stopping these rounds. Explosive reactive armor (ERA) is applied in appliqué plates to vulnerable areas, such as the vulnerable front glacis of main battle tanks. The plates are filled with layers of plastic explosive that is inert except under special conditions.

The armor is normally insensitive to small arms fire or random splinter hits, but in proximity to the shaped-charge blast of a HEAT (high explosive anti-tank) round or a KE long rod penetrator, it detonates and causes a counter-explosion that dissipates the kinetic energy of the strike.

Coupled with new glass composite or ceramic laminate armor, which also absorbs the energy of anti-tank munitions, this advancement reduces the destructive capability of once almost universally lethal rounds.

At the same time the tank cannon is approaching the performance limits for muzzle velocities and muzzle energies of conventional chemical propellant guns. This means that while tanks themselves might be more survivable against other tanks in future battles, they will be less survivable against smaller, cheaper, lighter, and more distant weapon systems like SADARM that can strike from afar and hunt them like falcons hunt rats.

Clearly, tanks and tank warfare will need to change dramatically to continue to be viable in regional combat situations against anything more than a fifth-rate enemy fielding outmoded weapons. Among other things, they will need to become faster and stealthier.

This means that they will need to become smarter, too, because already tank battles at just below the "double-nickel," the national speed limit, are the norm in armor-on-armor engagements, and such speeds are already almost too fast for the human nervous system to handle in warfare environments.

Better still, tomorrow's tanks will need to develop capabilities for sneaking up on enemy tank formations unseen and delivering their strikes from standoff distances, performing more like contemporary fighter planes in combat than participants in World War Two—era set-piece engagements.

The Dawn of the Hypertank

Ultimately, tomorrow's tankers will wage war in advanced, all-electric tanks equipped with revolutionary hypervelocity hyperkinetic weapons systems as their main armament and laser weapons as secondary weapons systems. Railguns, coilguns, or electrothermal guns capable of firing superdense metal ammunition at ultra-high velocities will replace current chemical propellant systems.

These main armament systems will not only be faster and deadlier than conventional tank cannon fire; they will be virtually silent, too. Computer targeted and automatically loaded, these hyperweapons will have the standoff range, speed, and agility necessary to engage fast-moving multiple targets at long distances, such as aircraft, UAVs, and terminally guided missiles. They will be the perfect armament for our stealthy hypertank.

Laser weapons will afford the hypertank with a second-line offensive capability against ground or smaller airborne targets, such as armored personnel carriers, other tanks, and helicopters or robotic UAVs. The laser weapons will also play defensive roles. As part of the tank's active defense system, they will dazzle the seeker heads of incoming projectiles or destroy them while in flight.

Other elements of the active defense system will use ultra-high frequency sound to defeat acoustic sensors of top-attack BAT (ballistic antitank) rounds and change the hypertank's thermal signature to defeat FLIR (forward-looking infrared) seekers.

The hypertank's powertrain and drive system, including steering, braking, and transmission, will be controlled and powered electrically. This approach would reduce its weight and eliminate design constraints on its interior configuration and crew carriage requirements by doing away with the bulky electromechanical drive and control mechanisms of its diesel-powered ancestor.

Advanced vetronics, a vehicle control interface, will allow the tank driver to maneuver using realtime video information coupled with computer-generated graphic overlays indicating position, threat analysis, and weapons management data, displayed on a lightweight head-mounted display. Networked global-mobile communications will transmit these data to battle managers at distant command centers. Such personnel could assist in targeting and even assume direct control of the hypertank in emergencies.

A number of technical problems will need to be

overcome before the hypertank rolls across tomorrow's battlefield. These include challenges in energy storage technology to power the tank's hyperweapons and the need to reduce the "Tempest" signature—the amount of electromagnetic energy that can be picked up on enemy sensors—that an all-electric tank generating large amounts of power will be likely to produce.

But the alternative to the hypertank is technological stalemate, and this state of affairs has never existed for long on the battlefield.

There Still Might Be Giants

There will still be some heavy metal dinosaurs prowling the fringes of tomorrow's battlezone and adding their growling din to the faint sounds of whispering death from lasers and hyperweapons.

These will be the battery weapons: heavy long-range offensive weapon systems like the MLRS (multiple launch rocket system) which can fire missile salvos across long ranges to obliterate enemy armor in a rain of hellfire and turn airfields, fortified positions, and other installations into blasted rubble.

Artillery guns of various types, including heavy-barreled howitzers of the M109 Paladin class capable of hurling a wide range of ammunition, from conventional dumb rounds to brilliant, tank-busting top-attack weapons, will also make their contribution to the carnage of war, aided by advanced targeting systems using satellite data to lock onto distant mobile targets miles over the horizon.

The Crusader self-propelled howitzer is tomorrow's version of such a system and is scheduled to enter service by the year 2005, according to post-QDR Army and Pentagon procurement plans. Crusader is to be the U.S. Army's first fully automated and computerized tracked combat vehicle system, and the first tracked vehicle designed for tomorrow's digital battlefield. The system, including the Future Armored Resupply Vehicle (FARV), is to replace the Paladin, which will be phased out as the twenty-first century opens.

Mobile Scud and SAM (surface-to-air missile) launchers, massive vehicles angling deadly offensive theater ballistic missiles and anti-aircraft rockets at the skies, will also lumber into place, waiting for the moment when sensitive tracking radars announce the moment to fire, then erupt in flames and smoke as missiles claw skyward on their errands of destruction.

Elsewhere in the theater, poised to stop these predators from reaching their targets, will be mobile defensive missile installations such as advanced block versions of Patriot and the newer Arrow systems. Patriot, the Scud-killer familiar from CNN coverage of Desert Storm, is a good example of these systems.

The Patriot Fire Unit (PFU) comprises radar, launchers, missiles, and battle management/command control and communications (BM/C3I) components. Radar finds and tracks the target while battle management and C3I engages the target and homes the missile in for the kill. Patriot is effective only against high-trajectory threats, such as Scuds, but not against low-flying threats such as cruise missiles, because current Patriot radars cannnot see over the horizon and identify long-range threats.

In tomorrow's war, command and control aircraft like AWACS may give Patriot an over-the-horizon capability by adding airborne surveillance and fire control to the missile system's ground-based sensors. A two-year series of experiments between 1994 and 1996 code-named ''Mountain Top'' have already shown this system workable against cruise missiles.

There will also be THEL—the tactical high energy laser system being jointly developed by the United States and Israel and using components of the Arrow ballistic theater missile defense system's Green Pine radar to provide targeting and fire control capability.

Using induced rotation laser technology to enable the lasing of more of the surface area of a missile in flight to increase destructive power, THEL batteries would be used against high-value targets, such as in-flight Scuds believed to be armed with chemical, biological, or nuclear warheads, and destroy them at the top of their trajectory.

Used in tandem with Patriot and other conventional anti-theater ballistic missile systems, THEL would greatly reduce collateral damage from falling post-contact debris that in the past has proved almost as deadly to both civilians and military forces as impact from Scud warheads.

Digital technology, hyperweapons, lasers, advanced tactical communications and battlefield sensors, and new types of small arms and man-portable systems will all play their roles in revolutionizing the land battle of the future. Many of these same advances will also reshape tomorrow's air war in remarkable ways. In the next chapter, we will examine what forms these transformational new developments are likely to take in the realm of tactical air.

SIX

TACAIR 2010

TOMORROW'S TACTICAL AIR (TACAIR) capability will be based on today's supreme military maxim: whichever side dominates the air war is the side that dominates the battlefield.

Air superiority clears the skies of enemy aircraft. Air superiority suppresses enemy air defenses (SEAD), sweeping the ground of highly mobile and short-dwell targets, including surface-to-air missile installations, tanks, and mobile rocket launchers.

Air superiority supports the safe arrival and resupply of friendly forces to the theater and protects defensive ground assets against attack.

Air superiority can be summed up in six words: first look, first shot, first kill. If a military force has got those three elements in its corner, it will most likely prevail. If it does not, it will most likely suffer stalemate or defeat.

On the strategic level, owning the air means possessing the capability to project military power globally, and to project this capability with great speed and survivability of forces. It means possessing the reach and the punch necessary to take out ballistic and cruise missiles and other weapons of mass destruction

(WMD) before they can be used by enemy forces.

Tomorrow's air force will be leaner and meaner, a more high-technology force optimized for the strategic role of fighting two simultaneous major regional conflicts (MRC).

It will be a force pursuing a policy of deterrence by denial, as opposed to deterrence by threat of punishment, as in the past. In true ninja fashion, it will seek to disarm opponents before they even know what hit them. Failing that, it will be capable of delivering withering strike power on enemy targets.

Tomorrow's air force will be a high-leverage force, one designed to support hard-charging surface assault elements as they storm an objective and secure sea-air-land bridges to the theater.

Tomorrow's regional threats will probably not resemble yesterday's or today's oppositional forces. Warplanners do not necessarily foresee the luxury of a lengthy Gulf War era–style buildup in the next MRC that the United States will have to fight. If U.S. and allied forces fail to bust in, hold fast, and quickly build up their force structures and supply lines, they might not prevail.

At the same time, the mainline fighter and bomber fleets of America and her Coalition partners have been approaching obsolescence.

While the F-15, F-16, and F/A-18 fighter planes still rule the skies, they are aircraft approaching the end of their useful service cycles.

With foreign warplanes equipped with sophisticated launch-and-leave missiles, these fighters are moving close to parity with the F-15. Advanced surface-to-air missiles (SAM), too, such as variants of the SA-8 Gecko and SA-9 Gaskin, are making today's formerly indomitable fighters more and more vulnerable to kills from ground-based threats. The vulnerability of the F-16 to these weapons was brought home by the 1995 downing of one such fighter over Bosnia by two Bosnian Serb SA-6 surface-to-air missiles, forcing its pilot to eject.

As for bombers, the world witnessed the continued ability of the battle-tested B-52 to drop tons of munitions on Iraqi ground forces during Desert Storm from high altitudes, and to do so with devastating effect on their ability to fight.

But technological advances in deployable missile defenses mean that tomorrow's enemy will have at his disposal cheaper, smarter, and longer-range ground-to-air weapons. With these, even the high-flying B-52 and its successor, the B-1B strategic bomber, will have become vulnerable targets.*

To maintain air superiority, the U.S. Air Force has invested heavily in new combat technology in order to maintain a flexible and responsive air strike capability that is also hard to defend against. In few other areas of military research and development has there emerged as impressive an array of superweapons which have captured the imaginations of friend and foe alike.

Few will forget the striking video images of a precision-guided munition dropped from an F-117A Nighthawk, plying the silent night skies over Baghdad, as it went straight down the airshaft of a government building and exploded with devastating force, or the fireworks-like displays of "triple-A" (automatic

---

*The B-1 flew its first combat missions during the four-day Desert Fox campaign of December 1998. The bomber operated out of air bases in the Gulf states of Bahrain.

antiaircraft) fire blindly directed at unseen U.S. planes as Desert Wind, the air strike phase of the Coalition attack, commenced. It was images like these that brought home the meaning of the much-bandied term "stealth."

Tomorrow's warplanes will continue to be stealthy. They will need to be, because tomorrow's enemy will be equipped with counterstealth radars, the better to see them and kill them.

They will be faster, too, as well as more maneuverable, and they will need these capabilities in addition to stealth, because tomorrow's missiles and opposition warplanes will also be faster and more maneuverable, just as they will be better equipped to spot their nimble game.

They will also be smarter, having advanced avionics, because at the supersonic and even hypersonic speeds of their operating envelopes, the human nervous system won't be able to maintain effective flight control, let alone fly the plane and target and shoot its weapons simultaneously.

The weapons systems the aircraft will carry will be smarter and deadlier, too, because the targets they will strike will be better defended, harder to hit, and located further away from the shorelines where friendly forces will land and concentrate.

As mentioned, some of the combat aircraft that will fight tomorrow's air war, such as the F-117A fighter-bomber and the B-1B bomber, are already in service.

These are being upgraded into new and more advanced versions. While this is happening, other aircraft, such as the B-2 Spirit of St. Louis and F-22 Raptor, are being rolled out.

Still other warplanes, such as the Joint Strike Fighter, are just beginning their development cycles.

The JSF, in its three planned configurations, will replace the F-16, the F/A-18 Hornet, and the Hawker Harrier for the British Royal Navy.

Nor is it just the United States that is engaged in building the aircraft for tomorrow's war, either. The former Soviet Union, the European Economic Community (EEC), and France, among others, are also engaged in new first-line fighter plane development.

Russia's famous Sukhoi Design Bureau and the Mikoyan-Gurevich (MiG) factories are at work on new, updated versions of the formidable SU and MiG fighter planes.

A consortium of EEC armament firms, under government contract, are currently developing the twenty-first-century Eurofighter. As for France, that country has already begun flight testing its own warplane of the next century, called the Rafale.

These advanced combat aircraft all have different strong points and weak points, and a comparison between them might very well take up an entire book in itself.

But in order to stand any chance of prevailing in the combat environment the aircraft are likely meet in tomorrow's war, they will all have to share four main characteristics—speed, maneuverability, advanced avionics, and low observability.

Low observable (LO) technology, otherwise known as stealth, is a combination of methods designed to defeat radar identification of combat aircraft by reducing that aircraft's RCS (radar cross section).

An RCS is produced whenever the beams of searching radars come into contact with aerodynamic surfaces on an aircraft's body, reflect off (backscatter), and are in turn picked up by receivers located on the ground or in the air.

When there is high backscatter of radar pulses, the aircraft shows up on radar scopes in high definition and is said to have a large RCS. When radar energy is absorbed and backscatter is dissipated, the RCS is correspondingly small and the plane is said to be low observable.

Reducing an aircraft's RCS affords it with a number of advantages over easier to see or "visible" aircraft. For one thing, it forces enemy radars to increase their transmission power, to emanate more strongly as they sweep the skies, making these radars more visible to friendly forces and easier to detect and destroy.

Low RCS also enables the design of special surfaces on the fuselage of planes designed to induce errors in enemy detection systems—they can, for example, spoof an incoming missile's homing systems into an inaccurate target picture by dazzling its radar or glinting its infrared trackers with thermal noise.

Further bonuses of low RCS include the reduction of the power of electronic countermeasures needed to jam enemy electronic warfare and to reduce the size of decoys and the quantity of radar-reflective chaff carried by any given warplane, thereby affording more room for offensive weapons payload. In short, a low RCS makes for an overall enhancement of platform survivability in a stress-filled combat environment, and gives the plane a better strike and defensive capability.

Sky Hammers: Tomorrow's Superplanes

Tomorrow's fighter and bomber aircraft will represent some of the most advanced aircraft ever developed for military or civilian roles. They will be the first of a new breed of warplanes. Ironically, it's possible they may also be the last.

First planned and developed during the Cold War to outfly and outflank the best that the Soviet Union could put into the air or launch from the ground, these high-technology aircraft have had to overcome major economic and political hurdles to justify their existence.

If global hostilities continue to scale back as they have done since the end of the Cold War, variants based on new aircraft designs could remain state-of-the-art for decades to come.

Although original force projections have been recalculated on the downside since the first Bottom Up Review in 1993, both the F-22 Raptor and the B-2 Spirit bomber have both survived the budgetary ax and will be contenders in tomorrow's major regional conflict.

They could possibly be joined by variants of the Joint Strike Fighter, or JSF, an aircraft still in the concept stage as of this writing. Unlike the JSF, the F-22 and B-2 have already passed the prototype stages and are now in active production. Yet unlike America's other superplane, the F-117A Nighthawk, neither has ever been flown in combat.[*]

These two unique aircraft designs embody two distinctly different yet complimentary visions of tomorrow's warplane. They provide glimpses into what the air war phase of tomorrow's MRC will be like, and how different it will be from that of past military conflicts.

---

[*] The B-2 flew its first combat mission in Allied Force as this book went to press.

F-22 Raptor: The Silver Bullet Plane

The most advanced production fighter aircraft in the world today is the F-22 Raptor air superiority fighter. The F-22 will replace the F-15 Eagle, which will be 31 years old by the year 2010 and will face challengers to the title of world's foremost air superiority fighter from advanced-design Russian and European aircraft.

Russian MiGs and Sukhois would be the F-15's chief global contenders, especially because of the former Soviet Union's shift from a policy of politically motivated arms sales to an aggressive marketing strategy in recent years.

The Russians today sell their best technology to virtually any buyer with the hard currency their economy needs to survive, regardless of political alignment, and the Russians show no signs of changing direction in future. Tomorrow's weapons systems and air defense systems will have also improved to the point where today's first-line fighter aircraft would have lost not only superiority, but even parity.

With the F-22 operational, the United States will continue to maintain the margin of technological advancement that will enable its air assets to prevail over any known or foreseeable challenge that may arise.

The F-22 multimission fighter plane embodies a blending of stealth, speed, and high maneuverability. It can draw first blood in an aerial dogfight as well as slip through radar defenses to deliver smart munitions to high-priority targets.

The Raptor is fast and agile enough to quickly outdistance and outfly pursuit aircraft. When cornered, it can stand on its tail and fly rings around other planes—and some missiles, too—that are chasing it through the skies. In short, its blend of design features and avionics give it a first look, first shoot, first kill capability bar none.

The QDR calls for a wing of thirty-eight F-22s to meet combat needs into the first decade of the next century. The small force of F-22 Raptors flying by 2010 will be the force-leveraging spearhead of a mixed air combat arm made up of advanced technology fighters, stealth aircraft, and last-generation planes like the F-16.

The Raptor force will perform the roles of an assortment of different aircraft. To quote from the USAF's vision statement, Global Engagement, the force is intended to "pack more lethal punch into a smaller package" than ever before.

And since the F-22 is likely to be the last major fighter plane developed along the lines of Cold War force requirements, it may be the last true superfighter ever produced in production quantities. Barring drastic returns to past superpower confrontation (which are unlikely), future generations of warplanes will not likely embody as radical improvements over the F-22 as the F-22 embodies over the F-15/F-16 generation of fighters.

Flight Performance Characteristics

The current generation of fighters achieve speeds just below Mach 1, the speed of sound, using standard military or "dry" power. To achieve multimach speeds, high performance turbofan engines have afterburners built into them. These spray fuel into the engine and burn it at a much faster rate, in the process supplying more powerful thrust.

Flying on afterburner consumes fuel at far higher rates than does normal thrust, which significantly reduces the time that a fighter can spend in the air. It also produces a great deal more heat energy, which increases the plane's radar and thermal signatures and makes it a more acquirable target to the homing radars and infrared seekers of air-to-air missiles.

The F-22's supercruise capability means it can cruise at supersonic speeds without afterburner in "dry" mode—supersonic speeds are therefore a standard part of the Raptor's normal operating envelope. Its twin Pratt & Whitney F119 engines have a far higher thrust-to-weight ratio than those found on previous-generation fighters. This means there is less inertia caused by the plane's mass for the engine to overcome in propelling the aircraft through the sky, making it far more agile than its predecessors.

The F-22's engines are also thrust-vectoring. The engine nozzles can be made to swing upward and downward in order to direct engine exhaust to enhance the power and agility of the plane during inflight maneuvers. The F-22's acceleration capability is approximately twice that of the F-14D Tomcat and F-15C and nearly a third better than the muscular SU-27. Its engines give it a twenty-five percent increase in combat range over the F-15.

It should be obvious from these statistics that the F-22's propulsion system makes it very well suited to high-g combat maneuvers involving sharp turns, dives, and wingovers. This has mandated certain advances in the design of the Raptor's fuselage to accommodate advancements to its power plant and propulsion systems.

The F-22's fuselage has a high titanium content—nearly 70 percent in the aft part of the plane—needed to withstand the tortuous gravitational stresses that the plane will endure during combat and to enable it to withstand the massive thrust potential of its engines.

Approximately 25 percent by weight of the aft fuselage comprises large electron-beam–welded titanium subassemblies called "booms."

The boom construction technique was devised to give the plane greater structural integrity and to improve stealth characteristics; the booms reduce the amount of rivets and welds in the airframe by about 75 percent over previous aircraft models. The fuselage was designed entirely on CATIA, a computed-aided design (CAD) system.

The F-22's wings, like the fuselage, were also designed using CAD technology. Each wing weighs approximately 2,000 pounds, made up of titanium, composite materials, aluminum, steel, and other materials in the form of fasteners, clips, and miscellaneous parts. Each wing measures sixteen feet by eighteen feet in length.

The wings are designed to cruise at supersonic speeds for extended periods of time and to withstand high-gravity maneuvers and are interchangeable from plane to plane. Unlike the wings of today's aircraft and some of tomorrow's fighter designs, they bear no weapons pylons, or "strakes." Weapon bay doors are conformal, dovetailing into the undersurface of the fuselage. Weapons carriage is internal, and a special dispenser lowers missiles into firing position once the bay doors are opened.

The F-22 is a stealthy fighter, far stealthier than any existing air superiority or air dominance fighter now flying. Its fuselage is a blend of the "curved" stealth of the B-2 bomber and the "angular" stealth of the F-117 Nighthawk fighter-bomber.

The Raptor is not, however, as stealthy an aircraft as either plane, but then again, it was never intended to be. When combined with its stealth characteristics, its high speed and formidable maneuverability make it hard to see and hard to hit even if seen, either by other advanced fighter aircraft or by the radar seekers of Russian-built SA-8 and higher surface-to-air missile installations.

Adding to innovations in the plane's overall flight performance will be an advanced avionics suite that will include innovative radars, ladars (laser-radar), and other new sensors.

The enhanced sensor capability will give the Raptor pilot greater situational awareness. He will track, identify, and fire at the threat before his opposite number can detect the F-22's presence. Sensors will provide information on type, location, speed, and direction of enemy aircraft, type and location of surface threats, effective range of enemy radar and weapons, targeting priorities and solutions, and all relevant information concerning the F-22's wingman and other friendly aircraft.

Sensors will deliver their information to digital high-resolution cockpit displays and the pilot will use advanced controls to fly the plane and fire its ordnance. Many systems currently requiring the pilot's attention will be automated, such as monitoring engine performance, hydraulics, fuel consumption, and the effects of g-forces on key structural elements and components of the airframe.

Mission data, including threat identification and navigational data, will be available to the pilot on a bank of multimode flat panel display screens capable of exhibiting high-density color graphics overlayed with alphanumeric combat data. All data will be in real-time and will be derived from several sources.

The principal source will be the Raptor's main active sensor array featuring a 32-bit processor based on PAVE PILLAR architecture, VHSIC modular circuitry, and ADA-based software. Other sources for the data fused together for the pilot will include airborne surveillance assets such as Joint Stars and AWACS aircraft, ground-based C4I stations, and military satellites in orbit around the earth.

The plane's onboard processor will fuse this data into a comprehensive picture of the combat environment. Threat envelopes from SAM sites, for example, would appear as graphical "balloons" surrounding high-definition pictorials of the ground below, while a projected flight path graphic would also overlay the imagery.

The conventional HUD or head-up display found in the F-15/F-16 fighter aircraft would be replaced with a lightweight head-mounted display (HMD) system similar to but more sophisticated than those currently worn by military helicopter pilots.

The HMD system will be a "HUD in a helmet" that will project mission data, including high-density graphics and pictorial overlays, directly in front of the pilot's eyes. Flight controls will be integrated into the HOTAS—hands on throttle and stick—system that will enable the pilot to control flight systems and weapons systems directly from the throttle. Armament of the Raptor is to include advanced airborne weapons, including the AMRAAM, AIM-9X, and JSOW missile systems.

The B-2 Spirit:
The Jewel in the TACAIR Crown

From its first clandestine test fight in 1982 to its official unveiling at Palmdale, California, in 1988, to is first successful test firing of a precision guided standoff munition in 1997, the B-2 advanced technology bomber has been perhaps the most secret of high-level aircraft projects.

At current production costs of some $2 billion per plane, the B-2 is easily one of the most expensive pieces of military hardware in the history of aviation warfare.

Yet despite years of controversy regarding its ability to live up to its advertised capabilities and a radical change in its original combat mission, the consensus in Congress and the Pentagon is to continue to fund the B-2.

Politically speaking, the B-2 continues to fly high while other advanced aircraft programs have been shot down in flames, making the B-2 a survivor before it ever sees combat. Post-QDR plans continue to be for twenty-one fully operational B-2 bombers to be delivered to the USAF's 509th Bomb Wing at Whiteman Air Force Base in Missouri—the same number of operational planes planned back in 1993 when Congress capped future acquisitions at twenty additional aircraft.

The B-2 had been called the ''jewel in the crown'' of U.S. TACAIR assets. In fact, the warplane is considered too valuable to fly into any combat environment where it might entail the risk of being downed by blind fire despite its stealth characteristics.

In 1990, then commander in chief of the Strategic

Air Command (since renamed Strategic Command, or STRATCOM) General John T. Chain told the House Armed Services Committee, ''I cannot see putting very many [B-2s] at risk during a conventional conflict if they were going to be exposed to high threat, dense type of activity in a small geographic area.''

This turned out to be an understatement—when Desert Wind, the air offensive over Iraq and Kuwait, commenced in January 1991, not a single B-2 was committed to the attack (or at least, no indication of this can be determined from a close study of open sources).

The less advanced F-117A Nighthawk and refitted Cold War–era B-52 bombers were deployed as the primary assets in bombing missions over Baghdad and strategic targets in the Iraqi desert. Only now, with the first successful firing of standoff weapons, does it appear more likely that the B-2 may be flown in combat when tomorrow's major regional conflict calls for a precision bombing mission that other aircraft are not able to handle.

Foreseeably, this mission might be one that calls for long-range penetration of enemy defenses and the delivery of ordnance on high-priority targets.

If it turned out that tomorrow's MRC—as considered likely—won't give Coalition forces much of a chance to build up theater forces and is coupled with an escalating global threat—such as a rogue state's possession of, and intention to fire, theater ballistic missiles—then using a B-2 force to deliver a preemptive tactical strike would be a viable option.

The B-2 combines the stealth and survivability of the F-117A Nighthawk with the range and payload of a B-52 strategic bomber. It is estimated that a force of six B-2s could put as much firepower on target as was delivered in the 1986 Libya bombing raid but launched from the continental U.S. instead of Europe with only a single in-flight refueling.

Such a strategic scenario comes close to the original mission for which the B-2 was developed during the late 1970s. The mission called for stealthy aircraft that could defeat Soviet radar coverage, penetrate deep into Soviet airspace, and deliver nuclear payloads on high countervalue targets, such as missile silos, nuclear weapons plants, hardened command and control bunkers, and major industrial centers.

Neither the aging B-52 nor the newer B-1B strategic bombers were considered to have the capabilities to fly such a mission. Neither plane was deemed stealthy enough or maneuverable enough to evade detection by new Soviet radar installations.

In the wake of the Soviet Union's collapse, critics claimed that the B-2's mission no longer existed and that plans for more than the additional two already delivered to the USAF should be scrapped.

These criticisms were bolstered by new revelations of design defects, including deficiencies in its radar cross section or RCS (meaning the aircraft was less stealthy than advertised), structural cracks in its aft deck, avionics malfunctions, including problems with critical terrain-following and terrain-mapping radars, and lack of software for various functions, including some weapons delivery.

Finally, there were no advanced standoff weapons, such as AMRAAM, available at the time, which would be necessary for the B-2 to function in its new post–Cold War mission as a force-multiplier in regional theater conflicts, taking out high-leverage targets and opening a corridor for ''visible'' warplanes like the B-52 or B-1B through which to fly follow-on bombing sorties. This would support a shift in strategy from the Cold War–era deterrence by threat of punishment to the contemporary strategy of deterrence by denial.

Since delivery of the first fully operational Block 30—which designates combat-ready planes—B-2 (AV-20) to the 509th Bomb Wing four years ago, most of the bomber's major functional problems have been addressed. By the start of tomorrow's war, the B-2 should be ready to fly its revamped mission as the spearhead of U.S. TACAIR elements.

The twentieth Block 30 B-2, delivered in May 1997, features a number of new enhancements, including full-capacity terrain-following/terrain-avoidance radar, improved navigation, all-weather operation, and weaponry enhancements, including carriage of the joint direct attack munition (JDAM).

By the year 2002, the U.S. should have a heavy bomber force comprised of sixty-six upgraded B-52H, ninety-five B-1 series, and twenty-one B-2 bombers. This adds up to a total of 182 of all three types in the inventory, enough aircraft to deal with the contingencies of two simultaneous MRCs and peacetime training requirements.

However, since the QDR, the role of the B-2 in fighting future wars has again come under fire by critics in Congress and elsewhere. This time, the "B-2 killer" argument ironically is focused on weapon systems whose earlier absence was cited as a reason for canceling the B-2 by earlier critics.

These weapon systems include AMRAAM and JSOW, cruise missiles whose fire-and-forget capabilities allow them to be fired from extreme standoff ranges outside enemy territory with high kill probabilities. The Global Hawk UAV, and other unmanned aerial vehicles under development, could also be used in similar roles when outfitted as weapons platforms. This being the case, critics are asking if such a development doesn't negate the B-2's core mission—to stealthily penetrate enemy airspace and fire or drop munitions at close range.

The B-2 Inside and Out

More than any other military aircraft, with the possible exception of the wasplike SR-71 spyplane, the B-2 Spirit seems the physical embodiment of stealth and predatory striking power. From its arrow-pointed nose to the sawtoothed trailing edges of its wings and aft fuselage, the black warplane aptly lives up to its name. The Spirit is every inch the ghost plane for tomorrow's ninja war.

A unique blending of curved and angular stealth characteristics lend the B-2 its distinctive shape. The warplane's fuselage, wings, and other aerodynamic surfaces are fashioned from radar-absorbant composite materials (RAM) and covered with about a ton of energy-dissipating ironball paint.

The air intakes and compressors of its twin GE-100 turbofan engines are mounted well back from the wings' leading edges, their W-shaped front configuration designed to break up radar returns—auxiliary intakes used on takeoff are located behind doors which shut once the plane is in the air to preserve the stealthy contours.

The troughs behind the intakes from which engine exhaust gases are emitted are wide and coated with a carbon-rich compound to reduce energy (the gases themselves are mixed with air to cool them and can be mixed with CFCs—chlorofluorocarbons—to eliminate contrails). The engines are specially baffled to reduce their acoustic signatures as well. The B-2 is as silent an aircraft as it is stealthy.

The plane's weapons bays are conformal, the hatch doors blending seamlessly with the undercarriage of the B-2 when closed. All upper surfaces are likewise blended together into a computer-designed whole, their individual parts dovetailing with the precision of a carpenter's mortise and tenon. The parts are joined using special classified high-technology processes that reduce seams to microscopic tolerances.

Combined with internal electronic warfare systems that jam, spoof, and otherwise counter enemy detection measures including, but not limited to, radar, the B-2 is said to be as hard to detect on enemy scopes as a bumblebee despite having a wingspan measuring almost two-thirds that of a football field.

Shrinking the B-2's radar cross section or RCS by the "shaping-and-masking" techniques described above is one of the primary elements of the plane's stealthiness. It formed the object of years of research and development, both when the Spirit was funded as a covert or black program, and after its official 1988 unveiling.

But using shaping-and-masking approaches to stealth design places certain performance penalties on the airframe of any aircraft depending on modified aerodynamic and control surfaces to lower its observability.

Stealthy cutout shapes such as those used in the B-2 and F-117 make the planes much harder to see on radars. But at the same time these same measures make them much harder to control by conventional electrohydraulic methods.

To compensate for the tradeoff between low RCS and inflight instability, the B-2 is equipped with an advanced suite of avionics including a computer-controlled fly-by-wire system that is hardened against the electromagnetic pulse (EMP) effects of nuclear explosions and which automatically compensates for the inherent instability of the airframe.

Another critical function performed by the B-2's advanced avionics suite is its enabling the aircraft with terrain-following/terrian-avoiding performance capabilities.

These features also contribute to the high stealth characteristics of the Spirit by allowing the B-2 to fly a ground-hugging "nap-of-the-earth" trajectory and enable it to get in beneath the downward limits of enemy radar coverage, or to fly "under the curtain," as the saying goes.

A View from the Flight Deck

All of the B-2's systems, including navigation, flight control, targeting and ordnance delivery, countermeasures detection, and management and secure communications, are accessed from the cockpit of the B-2. Virtually all the information the bomber's two-man crew is likely to need during a mission is displayed in color on four multiple-mode CRT screens.

The number of screens has increased from three since the first operational B-2 was delivered, the fourth screen replacing panels containing analog gauges.

Further study by the author of B-2 cockpit photographs taken between 1993 and 1997 shows a general trend to replacing analog gauges and pushbutton controls with digital screen technology. It's also possible that even more radical command interface technologies using virtual reality and interactive video (such as those discussed earlier in connection with the F-22 fighter) will be included in future versions of the plane.

The B-2's cockpit is spacious—the plane is sixty-nine feet long, approximately the same length as a typical New York City brownstone apartment—so the crew has room to stretch out during missions estimated to last in the vicinity of ten hours at a time.

In fact, there is room enough on the flight deck for a third crew station, which may be added in future blocks of B-2 design. Wraparound windows—two front windshield panels and two side panels, both incorporating energy-absorbent technology—give the crew members a panoramic view of the sky, and depending on flight conditions, of the ground below.

The windows do not appear to be removable, though, prompting questions about safely storing the aircraft in the baking heat of non–air conditioned desert hangars.

To the left of the crew cabin sits the pilot, whose tactical screens include navigation and flight control data and to whose right is a set of commercial airline-style throttles. To the right of the cabin sits the B-2's copilot/''Wizzo'' or weapons systems officer, whose front-mounted display panels reflect offensive weapons availability and targeting as well as electronic warfare and other ''soft kill'' capabilities.

Both crew stations are equipped with advanced design HOTAS (hands on throttle and stick) controllers studded with red and black ''pickle'' buttons enabling both crew members to carry out a wide range of control and offensive functions.

As alluded to above, the flight deck of tomorrow's B-2 might include the addition of what is officially termed an ILS, or integrated large screen display. Others call this the "god screen" concept. By any name, the ILS would be a large, integrated flat panel display screen in which all tactical and operational data are available to B-2 crew members.

The display system would do more than just depict ongoing events. The system would be responsive to eye, voice, and hand gesture commands, and it might include digital links to head-mounted displays (HMD), virtual reality interfaces that would enable all crew members to interact closely with and control all aspects of the warplane's performance and weapons systems.

The technology for such advanced control systems is available today—voice recognition, eye-trackers and datagloves, and the large flat panel displays, including the computer processors necessary to tie the system together, are all commodity items, and therefore are procurable as COTS (common-off-the-shelf) technology components.

Weapons Payload

The B-2 is capable of delivering both nuclear and conventional munitions of many varieties. With a payload capacity of about 40,000 pounds, it carries roughly the same amount of ordnance as the well-respected B-52, a plane justly known for its formidable saturation bombing capabilities.

Like the B-52, the B-2 is capable of dropping either dumb or laser-guided iron bombs at short distances from the target, but because of the revamped mission foreseen for the Spirit, that's not likely to be the mainstay of its battle participation.

In anticipation of new military threats by the USAF, the ordnance the B-2 will most frequently carry will be sophisticated long-range missiles, so-called "launch and leave" weapons, that can be fired from standoff ranges and strike distant targets. These would enable the B-2 to be miles from the launch point before unfriendlies realized what had happened.

One of the staples of the B-2's weapons inventory will probably be a variant of the AMRAAM or advanced medium range air-to-air missile, an air-breathing cruise missile that, much like the Tomahawk of Desert Storm fame, will fly a devious ground-hugging flight trajectory to deep penetration targets and destroy them with surgical accuracy.

Variants of the joint standoff weapon (JSOW) are also likely to be part of the Spirit's payload. The baseline model of this cruise missile design will carry combined effects submunitions—bomblets that separate from the main dispenser and strike preprogrammed ground targets.

Follow-on versions of JSOW will be equipped with a high-explosive BLU-108 single warhead and will be terminally controllable from the Wizzo's station by a camera mounted in the missile's nose. For those missions that call for free-fall weapons, the JDAM or joint direct attack munition will be available, the aerially delivered bomb's guidance system linked to GPS or the Spirit's own inertial navigation system (INS) for accurate targeting.

Finally, a recent announcement by the Department of Defense of the inclusion of a new tactical nuclear weapon in the U.S. inventory seems to indicate that the B-2s flying into tomorrow's combat zone may have unconventional weapons in their bombloads.

Although DOD would not officially link the new B61-11 nuke to the B-2, the characteristics of the compact low-yield nuclear device make it the perfect nuclear weapon for the Spirit to carry, were it ever deemed necessary in a tactical situation.

Other Tacair Assets in Tomorrow's Air War

THE JOINT STRIKE FIGHTER

Tomorrow's air war will see the combat debut of yet another superplane if forecasts being made by warplanners in the U.S., Great Britain, and perhaps other Coalition nations, such as Germany, are correct.

This is the Joint Strike Fighter (JSF), to be a multiservice aircraft whose international variants are poised to supersede or replace planes currently in service by the USAF, USMC, and British Royal Navy. These countries may be joined by other strategic partners in the multinational development program that is intended to produce three thousand operational aircraft by early next century.

Whether any JSF aircraft will actually be in service by the year 2010 is debatable, though. Unlike the F-22, the B-2, or the F-117, the JSF has been flown only in computer-generated simulations where it was a contender in joint strike warfare campaigns taking place in MRC scenarios staged "in the 2010 time frame," as a recent DOD (U.S. Department of Defense) white paper puts it.

Physically, the JSF currently exists in the concept phase only, in the form of an assortment of prototype systems that are intended to coalesce into a fieldable aircraft in the near future.

Exactly when rollout will occur is a conjectural matter at this stage of JSF development. The U.S. Department of Defense claims that a JSF demonstrator will be capable of flying in combat in the year 2001, but a look at the development cycles of other new TACAIR initiatives, such as the ATF and stealth programs, cast serious doubt on this projection. The B-2 Spirit, for example, had been around for considerably more than a decade before it flew.

Nevertheless, an aircraft like JSF will be necessary to fill fighter-bomber roles currently served by aging F-16s, A-10s, and to some extent even B-52 heavy bombers (strategic aircraft deployed tactically during Desert Storm), when tomorrow's next regional conflict erupts. In two cases, the survivability and lethality of the above-named aircraft should by then be seriously threatened by parity with other aircraft and vulnerability to next-generation SAMs; in the case of the A-10, that plane should no longer be in service.

In general appearance and aerodynamic characteristics—and probably in performance, too, ultimately— the JSF bears an uncanny resemblance to the YF-23 ATF prototype, one of the two original contenders for the advanced tactical fighter (ATF) program of the 1980s. Demonstrator testing of both designs culminated in the selection of the YF-22 for development by the USAF.

The YF-23 has a larger, more muscular airframe than the F-22, a more powerful ramjet engine, and a more blended stealth configuration than the somewhat more angularly stealthy F-22.

During the ATF trials it became apparent that the YF-23 was an aircraft optimized for long-range missions, somewhat less agile and less maneuverable, and generally less of a "hot dog" plane than the F-22. These aspects made the YF-23 design less attractive to the Air Force as a plane that would replace the F-15, an air-superiority fighter optimized for the dogfight role, not an air-dominance fighter like the F-16, whose missions included deep-strike bombing runs.

Though the YF-23 was canceled, it was apparent that the worthy design of the aircraft was destined to be resurrected when an F-16 successor was sought, and this now seems to be the case. The factors that doomed the YF-23 to cancellation in the 1980s currently make the plane's design characteristics, transitioned to the JSF, attractive as a replacement for the F-16, A-10, A-67, and F/A-18A.

Another canceled ATF plane project whose technology base was transitioned into JSF that bears mentioning here is the former Navy AX program. The AX was to have been a low-observable multirole fighter-bomber whose airframe was based on a radical flying wing design. The AX would have been optimized for deep-strike and antisurface warfare roles and capable of carrying all existing USN missiles except for the Phoenix AAM.

The JSF will fly missions that require endurance, range, penetration of air defenses, and delivery of bombloads per sortie in the medium bomber ballpark. It's to be a plane that can hit targets far inland when launched from carrier flight decks or from more distant land bases separated from the theater by hundreds—or, with air-to-air refueling (AAR), thousands—of miles.

The high production numbers projected for the JSF is another factor making the plane a viable component of tomorrow's combined arms battlefield, as is the

JSF's planned performance envelope. The current roster of next-generation strike aircraft are "crown jewel" assets. Their high production and maintenance costs have been justified by their ability to attack successfully well-defended, high-value targets that could not be effectively engaged by any other means.

But conducting a successful air campaign requires more than just a precision scalpel; it also takes the cheap and dirty sledge hammer of round-the-clock saturation bombing to attack unfriendly infrastructure on a massive scale and disrupt enemy morale and will to fight.

This means having a large enough number of aircraft continually flying a large enough number of mission sorties to get the job done. Less expensive to build and maintain than the F-22, more flexible in the types of missions that it can fly, the JSF will be a day-night, all-weather warplane that delivers maximum bang for the military buck.

One of the reasons that the JSF will be less expensive will be its modular design approach, intended to produce variants based on a basic technology set. Once the core technology has matured, all variants of the JSF could then be assembled on a single production line.

Use of composite materials and off-the-shelf (COTS) subsystem components are other cost reduction measures incorporated into the JSF design philosophy. Modular and upgradable, yet technologically advanced, the JSF will be inexpensive enough to field in relatively large numbers, if current plans materialize along predicted lines.

The JSF will be built in three basic variants, two naval versions and an Air Force version. The USN wants a survivable first-day-of-war strike fighter aircraft to complement its current first-line fighter, the F/A18E/F. The USAF wants a multirole fighter plane to replace the F-16 and the A-10, the latter to be retired.

The USMC and the United Kingdom Royal Navy want STOVL (short takeoff/vertical landing) versions that can lift vertically from the flight decks of aircraft carriers and other ships capable of carrying aircraft. Currently, the only production aircraft that can perform this feat is the Sea Harrier, which is in service with the Royal Navy and also scheduled to be replaced with one of the JSF variants.

If successful, a STOVL-capable JSF would lessen the need for large flat-top aircraft carriers with long flight decks and the elaborate system of steam catapults and arresting pendant assemblies that are currently necessary for the effective takeoff and landing of naval warplanes at sea.

This isn't to suggest that a STOVL JSF would spell the end of the modern nuclear aircraft carrier. It would not, at least in the foreseeable future. What it would do, however, is make feasible the development of stealthier, cheaper, and perhaps even faster, but definitely *smaller* types of aircraft-carrying ships that could be used to spearhead a global strike capability in line with the Navy's Copernicus vision statements.

According to the Joint Initial Requirements Documents released by the U.S. Defense Department (JIRD-I and JIRD-II) the JSF will be a plane that can take a beating, yet continue performing over and over again. Its "enhanced sortie generation rate" (SGR), says the JIRD, "can deliver impressive combat power as a force application tool available to the Joint Force Commander (JFC)."

This document goes on to state that the JSF will

have a "lean logistics footprint." This means that JSF will be optimized to fly more sorties than present-day fighters of its class, with less turn-around time per mission. It also means it will be able to carry a larger and more sophisticated weapons load than preceding generation fighters.

The JSF will be sent out in the initial phases of tomorrow's MRC. It will be among the first planes in and among the last planes out of the combat theater. Its primary mission will be to kill short-dwell and mobile targets such as tanks, APCs, towed artillery, and mobile Scud launchers, and also to slam fixed tactical targets such as hardened communications nodes and aircraft hangars, supply depots, radar installations, and the like.

Desert Storm and subsequent wargaming analysis have demonstrated that suppression of this set of targets is critical to a quick and decisive victory in a future MRC. Studies conducted by the Defense Intelligence Agency (DIA) calculate that a range of approximately 400 nautical miles (NM) into enemy territory will enable the JSF to strike about 90 percent of its designated targets. Naval versions of the JSF set the range at between about 450 and 550 NM, while the USAF version will carry enough fuel to take it out to 600 NM without requiring refueling in flight.

The new joint standoff weapon (JSOW), an air-launched cruise missile that is effective in the mission the JSF is intended to fly, is the primary weapon system that the JSF will carry in its internal bay. Other weapons systems the JSF may deploy will be the JDAM (joint direct attack munition) and various general purpose bombs, mines, and rockets, such as the projected two 2,000-pound laser guided bombs that each JSF will be able to carry.

The JSF's avionics system will most likely incorporate an integrated off-board information management system for the pilot in the form of an integrated helmet-based head-mounted display (HMD) that would be responsive to voice, brainwave, or touch commands, or a mixture of some or all of these control interface command sets.

Preliminary tests have indicated that such an advanced pilot control system for the JSF may be less expensive to field than conventional HUD and stick systems and also would afford better control characteristics to the pilot. The IHAVS system, an integral helmet audio-visual system, has already been tested in JSF proof-of-concept demonstrations and has proved superior to conventional HUDs—head-up displays—found in current-generation fighter planes such as the F-15, the F-16, and the F/A-18.

Despite inevitable delays that are certain to hamper development of the JSF program, the plane seems to have broad support at the Pentagon and in Congress, which has kept the program alive in the QDR while other programs have been axed. Whether or not the JSF will fly by the year 2010 depends on how well the JSF program produces before the next QDR in 2002 takes another look at its viability.

F-117 NIGHTHAWK

Of all the advanced tactical and strategic warplanes that have been developed thus far, the F-117A "Stealth fighter" is the only one to have flown in combat during the Desert Wind air phase of the Gulf War (and again, the word to keep in mind is "officially," since we have no way of knowing with absolute certainty which aircraft were and were not flown in any given stealth mission).*

The faceted or angular stealth of the Nighthawk began with the secret DARPA-run Have Blue program in the 1970s and resulted in the spectacular surgical bombing missions of the Desert Wind air war over Iraq.

Actual nose-camera images from laser guided bombs in flight to their targets released to the news media demonstrated the ability of the Stealth fighter to deliver its ordnance with pinpoint accuracy. The black ninja planes had flown into Iraqi airspace and into the center of downtown Baghdad undetected by enemy threat-seeking radars.

All the same, and for similar reasons to those cited in connection with the B-2, the F-117A is a hard plane to maneuver in flight. The strange geometries of the aircraft's aerodynamic control surfaces, the mainstay of its angular stealthiness, make it inherently unstable when flown.

In fact, without its sophisticated, computer-controlled inertial navigation system (INS), the Stealth fighter would get about as far as a broken kite in a high wind. Even with this computerized augmentation, the plane is notoriously difficult to handle. Pilots who have flown the F-117 refer to the Stealth by the less-than-honorific title of the "Wobbly Goblin."

In 1996, an F-117A attached to the 49th Fighter Wing crashed on the Zuni Indian Reservation in New Mexico, an area populated by members of the Pueblo and Zuni tribes. While the crash is the first publicly acknowledged incident of its kind, it is not the first time an F-117A has experienced serious in-flight control problems, at least while on test or training runs.

As of this writing, all components of the F-117A

---

*See note on page 120.

fleet are scheduled to have their INS systems upgraded with an improved version promising greater
aerodynamic stability for the aircraft. This is the Ring
Laser Gyro/Global Positioning System Navigation
Improvement Program system or RNIP-Plus retrofit.

RNIP-Plus is designed to integrate the F-117A's
INS with GPS data from the array of twenty-four
global positioning satellites currently in orbit around
the earth. The system is said by test pilots who've
used it to reduce significantly the aircraft's tendency
to drift off course. RNIP-Plus also increases targeting
precision, aligning target crosshairs to more precise
calibrations during missile launch and bomb delivery.

Because of maneuverability and other issues concerning the F-117's design features, the Nighthawk
will strike only under the cloak of darkness. In daylight combat operations, the F-117A would become
extremely vulnerable even to mediocre enemy fighter
planes chancing to sight it visually, although it would
continue to remain invisible or extremely hard to see
by other means, such as on airborne search radars.

Once sighted, though, it would become an easy target of opportunity. The Stealth fighter is not fast
enough, nor is it maneuverable enough, to escape a
true combat fighter, and its air-to-air ordnance is virtually nil. The Stealth is optimized for covert offensive strike missions and for little else.

This is one reason why the Stealth will fly only
night combat missions and will be used only against
high-value targets whose destruction merits the hazard
of exposing any one of the fleet of multimillion dollar
aircraft to any form of danger, however slight or in-
tangible that risk might be.

But on those missions that it is cleared to take on,
the Nighthawk will continue to prove an effective pre-

cision instrument—like its namesake, this stealth
plane can swoop in and rid the landscape of hidden
vipers, and open the doors to chain-lightning sorties
by faster and more heavily armed aircraft to hammer
the enemy into dazed submission.

Foreign Contenders

The United States is not the only nation engaged in
the development of advanced tactical fighter planes
optimized for air combat missions in the next century.
There are several ongoing initiatives managed by foreign nations, including Russia, France, and Sweden,
and also by European consortia including the forenamed countries, along with Britain and Germany, to
produce new combat fighter aircraft.

Variants of the famous Russian MiG and Sukhoi
fighters will surely be participants in any future war
large or small, especially since the former Soviets
have increasingly turned to foreign arms sales as a
means of raising vitally needed hard currency.

Today's practice by the Russians of hybridizing
their planes with a mix of native and Western flight
control and weapons systems to tailor aircraft to buyers' needs makes them extremely cost-effective pur-
chases for Third World powers. For this reason alone,
MiGs are destined to become a major component of
the air arms of undeveloped and nonaligned nations,
and therefore the most prevalent aircraft the U.S. and
her allies are destined to meet in air combat during
future confrontations.

The French Mirage 2000 is another warplane that
will be likely to fly in tomorrow's air combat theater,
in the hands of friendlies and adversaries alike. The

Mirage 2000 is a multirole air superiority fighter capable of being fitted with sensitive radars and of carrying a wide variety of missiles, including AMRAAM. The newest version of the plane is the Mirage 2000-5, capable of tracking twenty-four simultaneous targets, and computing simultaneous missile firing solutions for four of these. The first export models of this plane were delivered in 1997.

The French, however, will also have the new Rafale, an advanced tactical fighter being developed in air force and naval variants and designed for service as an air superiority fighter in the F-22 performance class. There are to be three variants, single- and double-seat air versions (Rafale B and Rafale C), and a single-seat naval version (Rafale M).

The Rafale will feature twin high-performance ramjet engines capable of cruising at better than Mach 1.2 speeds in "dry" mode—that is, without the use of afterburner. Its airframe will feature variable geometry canards—forewings mounted behind the nose to optimize aerodynamic efficiency and stability—and a fly-by-wire control system for high maneuverability. The Rafale will also incorporate low-observable (LO) airframe characteristics to make it stealthy to radar and infrared sensor systems.

Weapons for the Rafale will include French MICA air-to-air missiles and the capability to carry the French-developed ASMP nuclear cruise missile. The Rafale's avionics suite will be controlled by a head-mounted display instead of the conventional HUD, a development similar to that projected for the F-22 and JSF; the French have a highly developed technology base in the military electronics sector, and all navigation and weapons engagement and launch systems are likely to be very good.

The Eurofighter is another advanced tactical fighter development program that has been ongoing for several years. The Eurofighter is being developed by a partnership including the United Kingdom, Germany, Italy, and Spain, and is intended to produce air force and naval variants of the plane for service in all the participating countries.

Like the Rafale, the Eurofighter's airframe features front wing canards, an advanced avionics package, and supercruise capability without afterburner. Unlike Rafale, the Eurofighter has been the subject of many production delays and may not enter service until after the year 2010 if these problems continue.

The JAS-39 Gripen is a product of the Swedish JAS industrial groups and a plane designed both for national service and for foreign export. The Gripen is an air-superiority fighter, which, like the two other European planes already mentioned, uses front-mounted canards for maneuverability and stability. The Gripen is not, however, as ambitious a tactical fighter as either the Rafale or the Eurofighter, but it does have some unique qualities that the others lack.

Its small size and takeoff weight optimizes the plane for STOL (short takeoff and landing) situations such as non-military runways and roads, and short mission turnaround times of ten to twenty minutes. The Gripen's armament package will include AIM-9L Sidewinders and AIM-120 AMRAAM missiles, giving it an air-to-air and air-to-ground capability matching those of the world's first-line fighter planes.

The Russian Stealth Contender

By now the perceptive reader should be asking himself an obvious question—what about the Russians? If the likes of Sweden are in the advanced tactical fighter business, is it credible that the Bear has nothing more than souped-up versions of MiGs and SUs in store for tomorrow's air warfare?

The answer is that of course, the Russians are very busily at work in developing a new generation of stealthy warplanes to match those possessed by the United States. In fact, the general designer for the Mikoyan-Gurevich (MiG) aircraft bureau has publicly stated that "any aircraft delivered around the end of the century that does not fully incorporate stealth does not have a chance of success."

Russian ATF and stealth programs are highly classified. However, some information on where they are currently heading has leaked through the former Soviets' security net. While there are programs to match each of the major stealth aircraft systems of the U.S., such as a Russian Stealth bomber or "B-2sky," as some disparagingly refer to it, the pinnacle of Russian warplane R&D is their next-generation tactical fighter plane, the MiG 1.42 multirole fighter, or MRF.

Like its Western counterparts, such as the F-22, the MRF will incorporate two-dimensional thrust-vectored 180kN Soyuz R-79 engines and aerodynamic control surfaces that are optimized for low-observability.

Also like U.S. stealth planes, the MRF is designed for internal weapons carriage. Its principal weapon is a Russian AMRAAM clone, the Vympel R-77 medium-range AAM. The MRF also will use the Phazotron Zhuk-PH radar, a compact and powerful radar suite that incorporates highly developed synthetic aperture (SAR) features, such as terrain avoidance and moving target acquisition modalities, and is said to be in the performance ballpark of Western counterparts—a Russian first, if true.

Some reports state that the MRF has already flown; others say that it is now nearing flight testing. What is certain is that budgetary constraints have made the Russian ATF program flow at a snail's pace compared to similar U.S. efforts.

It's nevertheless impossible to be fully certain of just how far along the Russians are in the MRF program, which is clearly a top priority for them in order to maintain parity with advancements in Western fighter design that would turn their current planes into yesterday's technology in tomorrow's war, and deal yet another blow to national pride.

Tomorrow's Helicopter Gunships

Prowling the lowest tactical envelope of the air war, thirty, forty, or fifty feet off the ground, helicopter gunships will contribute their formidable firepower to the accelerated tempo of tomorrow's battlespace. Helicopter gunships will support infantry, kill tanks, hunt mobile and stationary Scud launchers, identify targets of many varieties, and perform a range of other gritty "down-in-the-weeds" jobs.

Because of the usefulness of helicopters and the comparative simplicity of the technologies necessary to build them, tomorrow's war will see many strong contenders in this area of procurement and development.

Tomorrow's combat choppers will be stealthier than yesterday's helicopters, with both a reduced radar cross section and a reduced audial signature. They will be better armed as well and more responsive to pilot commands due to advanced onboard avionics systems.

Of the many combat helicopter development programs that have been started in the last ten years, by far the most revolutionary has been the U.S. Army's LHX (Light Helicopter, Experimental) program, whose goal was to develop an advanced combat rotorcraft that could perform missions that other combat choppers, like Apache or Cobra, could not do and to go places they could not go. The end result of the program is the current RAH-66 Commanche helo.

The Commanche was originally planned to have a light transport version in addition to scout/attack variants, but successive budgetary cuts have done away with the former and have whittled down the final production numbers to an approximate 120 production models in the QDR.

The Commanche's main role is to be armored reconnaissance. The helo's mission is to seek out and destroy enemy forces and to designate targets for other weapons systems, primarily the Apache combat helicopter. It will also spot for artillery and fighter aircraft higher up in the envelope.

To help it perform its mission, the Commanche has a low-observable design that includes carriage of weapons internally and radar deflecting aerodynamic surfaces similar to those found on stealth warplanes. The doors of the Commanche's internal weapons bay can also double as defensive weapons platforms with three hardpoints each for missiles such as Hellfire or Stinger, or it can alternatively be mounted with rocket pods.

For attack weapons or auxiliary fuel tanks for long-range missions, stub wings can be added onto the sides of the chopper. The wings can support a maximum Hellfire missile payload of eight missiles, four under each in Apache-type two-plus-two racks.

With room for an additional six missiles in its internal bay, this gives the Commanche a total weaponload for the Hellfire of fourteen rounds. The Commanche can also be adapted to accept other weapon mixes as well, and new weapons types as they become available.

The main gun for the Commanche is the GE Vulcan II twin-barrel 20 mm machine gun. The lightweight electric-powered chain gun is housed in a special trainable turret developed by GIAT industries, a French weapons consortium. The Vulcan's gun barrels can be shrouded by a radar absorbent fairing for added stealth characteristics when entering heavily defended airspace, although this means an extra weight penalty and consequently a possible reduction in weapon payload per mission.

The Commanche features advanced avionics integration and logistics support integration, meaning that its weapons and flight control systems are all electronically assisted.

A wide-field-of-view HMD will incorporate realtime video, forward-looking infrared (FLIR), and digital map display overlays piped through a high-speed data bus of very high-speed integrated circuit-(VHSIC) based computers. The electro-optical Aided Target Acquisition/Designation (ATA/D) and Longbow Night Vision Piloting Systems (NVPS) will also be part of the Commanche's avionics package.

The main onboard communications system of the Commanche will be the ICNIAS lightweight integrated communications, navigation, and identification avionics system. Subsystems under ICNIAS will include

SINCGARS and HAVE QUICK 2A radios, a GPS satellite navigation system, and the HF SBBV, a location and identification system utilizing the Mk12 antijam IFF (identification friend or foe) transponder.

The six-year-long Commanche development program ranks as one of the longest in Army aviation history, but its results in the fielding of what is probably the most advanced helicopter design ever flown will undoubtedly pay large dividends in ensuring battlefield supremacy for friendly forces in tomorrow's MRC.

Friendly infantry and tank formations will especially look on Commanche as a low-altitude guardian angel, while their enemy counterparts will surely see it as a nightmare draped in black.

Spies in the Skies

Aerial reconnaissance, target identification, battle damage assessments, friend or foe identification, electronic warfare, and other related missions require another class of aircraft that can be categorized together under the general heading of "spies in the skies."

The uniqueness and special performance characteristics of these aircraft lie less in their speed or maneuverability or in the radical design elements of their airframes as in the data processing and/or information warfare payloads that they carry.

These systems enable the aircraft to perform missions that are critical to the favorable outcome of the battle, and as such they are every bit as indispensable as the advanced tactical fighter, the Stealth bomber or the ground-pounding helicopter gunship.

JSTARS and AWACS

JSTARS (joint strategic tactical airborne radar system) and AWACS (airborne warning and control system) aircraft perform complimentary functions. Both planes are aerial surveillance platforms that relay and coordinate information on and to shooters in theater, but concentrate on different parts of the battlespace.

JSTARS looks down at things happening on the ground. Its long-range air-to-ground surveillance systems are designed to locate, classify, and track multiple ground targets in daylight and darkness and under all weather conditions. Tank formations, supply convoys, airfields, troop movements, SAM sites, and Scud launchers are all part of what JSTARS sees.

The entire JSTARS system consists of two components, one airborne, the other located on the ground. The airborne component is the E-8C, a modified Boeing 707 equipped with a phased array radar in a twenty-six-foot-long "canoe" fairing located beneath the forward portion of the fuselage.

The computer-processed radar, which has a range of more than 120 miles, operates in synthetic aperture (SAR) mode, which provides extremely high-quality displays of targets under surveillance. The ground component is made up of multiple GSMs or ground station modules, which are five-ton trucks and other vehicles, such as HMMWVs (High Mobility Military Warfare Vehicle, also called Hummers or Humvees) containing downlink computers. Together, the system can provide real-time, wide-area tactical information to air and ground commanders.

Conversely, AWACS scans the skies, looking outward in all directions at friendly and enemy aircraft in the air combat theater. The main component of the system is the Westinghouse APY series radar, which is rated as the most powerful and versatile of any airborne radar in use today. The radar is located in a broad, flat, circular rotodome that sits atop a twin-strut pylon rising from the aft section of an E-3 Sentry aircraft, another Boeing 707 variant, directly behind the aircraft's wings.

As the rotodome rotates, the radars continuously sweep the skies for low- or high-level targets out to approximately 250 miles in all directions. Radar signals are fed down into the battery of computers linked to multimode display consoles within the fuselage of the plane which are monitored by a crew of airborne warning and control operators.

As the computers process the signals, they enable AWACS to track and identify fighter planes, bombers, tanker aircraft, unmanned aerial vehicles, missiles, and other airborne combat assets—virtually anything that moves through the skies and is likely to pose a threat to friendly forces. AWACS can also conduct electronic warfare missions using a suite of electronic countermeasures (ECM) detection and jamming systems.

The F-111E and Other EW Aircraft: Of Ferrets and Ravens

Electronic and information warfare aircraft are known as "ferrets" and electronic warfare (EW) itself is called, by some, "raven" warfare, and its human practitioners "ravens." The symbolism is pretty clear-cut. Ravens fly in the dead of night and the aircraft ferret out electronic emissions of enemy radar installations and other command and control systems.

These emissions go under the general classification of "Tempest," and they can be categorized and analyzed to determine the specific signature of different radars, communications systems, and any other systems which electronically process information of any type.

Once the Tempest data has been collected and analyzed, EW aircraft can then take one of two basic forms of action. The first is to launch electronic countermeasures (ECM) to jam, spoof, or otherwise interfere with the enemy's electronic information systems.

If these systems respond by defensive electronic counter-countermeasures (ECCM), then the ferret aircraft can either retaliate with stronger ECCM or opt to use more lethal offensive measures.

These include radar-homing (HARM, for high-speed anti-radar) missiles, beam-riding weapons that can lock onto signals broadcast by radar antennae and home in on the target installation, blowing it sky-high with warheads containing thousands of pounds of high explosive.

The ferret aircraft can either launch HARM rounds itself or hand off this duty to other airborne assets, including fighter planes and helicopters equipped with HARM weaponloads. This is one reason why the F-111E is frequently found operating in tandem with other aircraft during combat missions.

As the F-16 warplane ages and is replaced by the Joint Strike Fighter, it may find a new role among the ranks of ferret aircraft. When outfitted with electronic warfare pods, the F-16 is considered by many military analysts to be well suited to the EW role in the twenty-first-century combat theater.

Satellite Warfare

At the uppermost segment of the air combat envelope are found the military's space-based assets. In
the combat theater of 2010 to 2020, these will probably continue to be military and intelligence satellites
such as the ones in use today and their next-generation
variants and descendants.

Satellites have become critical to the successful
prosecution of present-day theater and strategic military engagements, and they will become even more
critical in the future where more complex logistics
and communications infrastructure will be necessary
to the operation of high-technology battle systems.

But while today orbital assets are still able to perform their missions without hindrance from the offensive weapon capabilities of unfriendly powers,
tomorrow's battlefield dynamics will make them increasingly vulnerable to attack.

There have been numerous programs, both black or
covert and white or open, to develop and test antisatellite weaponry capable of damaging or destroying
orbital assets without which important military communications and intelligence linkages quickly fall
apart.

Northrop-Grumman Corporation, a Department of
Defense contractor, announced on October 7, 1997,
that the AN/AAQ-24 Directed Infrared Countermeasures (DIRCM) system had successfully completed its
first airborne trial against a simulated missile threat.
The testbed was a British Sea King helicopter
equipped with DIRCM, which was developed for the
British Ministry of Defense (MOD) and the U.S. Special Operations Command (USSOCOM).

Meanwhile, the USAF has been conducting tests using its airborne laser (ABL) system. The system is built around a 10,000 watt chemical oxygen iodine laser (COIL) coupled with a deformable mirror system that compensates for atmospheric distortion of the laser beam and a beamwalk mirror system to steer the beam to its target. ABL would be mounted on the nose assemblies of 747-400F cargo planes and is scheduled to be fielded in a proof-of-concept test in 2002—that is, shortly after the next QDR.

There is also BLAM, the barrel-launched adaptive munition, which, like the above program, could be utilized as either an antisatellite or an antimissile system, capable of killing Scuds and orbital spy platforms alike.

The BLAM system concept is built around a cone-shaped, five-inch nonexplosive round or flechette (dart) with a needle nose that would be fired from hyperguns carried on high-flying aircraft, such as an F-22 fighter or B-1B bomber. Special "tendons" near the flechette's nose would steer BLAM rounds toward their targets, compensating for evasive tactics and atmospheric turbulence.

As is the case with land-based hyperkinetic weapons, BLAMs would derive their killing power from the transmission of their high kinetic energies to the target, resulting in penetration, explosion, melting, and fragmentation.

For killing satellites, BLAMs would need to travel at approximately 6,000 feet per second or better; for antimissile defense, about 4,000 feet per second would be necessary. An air-to-ground version would also be available for antiarmor roles with the Specter AC-130 gunship (a plane, not a helicopter) or the

Commanche helicopter as possible platforms with a range of fifty miles or better.

The U.S. Army is also developing a hyperkinetic-hypervelocity antisatellite system called the kinetic energy antisatellite system (KEAS). This is to be a ground-based system of satellite tracking sensors linked to hyperguns capable of firing KE projectiles into orbital space to destroy unfriendly surveillance satellites.

Each of the forenamed systems would have potential in antisatellite warfare roles, especially ABL, which appears to be a more powerful system designed for deployment at higher altitudes, which are more conducive to laser warfare due to the thinner and less distorting atmosphere that far up, close to the leading edge of space.

Within the approximate eleven-year time frame between today's fragile peace and tomorrow's projected war, it's certain that anti-satellite weapons technology will have improved to the point where satellites can be targeted by lasers based on the ground, carried on planes, or even placed in orbit around the earth.

Coupled with the fact that war games conducted by both government and private agencies have demonstrated an inclination among combatants armed with deployable anti-satellite weapons to use them in the first hours of a future war, this spells trouble for space-based military assets in their present unprotected forms.

Clearly, by tomorrow's MRC, American and Coalition satellites will need to be far better protected against enemy countermeasures. If not, then friendly warfighting capabilities and the kind of quick soft-kill and winnable war that planners anticipate will be in serious jeopardy of turning into something far more costly, painful, and sanguinary.

And while air power will reshape the future battle-field, as it has been doing since the dawn of aviation warfare, and space power will add a third dimension to air combat operations both in offense and defense, ultimate victory in tomorrow's war, as in today's international conflicts, will continue to hinge on mastery of the planet's great oceans. In this context, we will next examine the role that naval power will play in the opening decades of the twenty-first century.

SEVEN

• • • • • • • • •

From the Sea

0320 HOURS.

Dawn's light has not yet reached the littoral waters of southeastern Iran. Sea, land, and sky remain shrouded in darkness. Nothing seems alive or awake. Yet in the moonless blackness, observers positioned behind barbed wire emplacements along a six-mile strip of beach scan the ocean through infrared night observation devices.

Though they see no sign of the enemy, the men on the beach are tense with grim expectation. They know that seaborne attack is imminent, and they know that the beach presents a probable choice for a landing by U.S. troops and America's Coalition allies.

Further inland, thirty feet below ground in a bunkered command center, combat-fatigued scope men nervously scan banks of radar screens. The radar systems are of current Russian design, linked to high-speed computers that constantly comb through the data stream of remote telemetry, sifting for the telltale threat signatures of low-observable weapons platforms.

They are new counterstealth radars, touted by the Russians as a guarantee against surprise attack by advanced stealth warplanes and ships. The scopes con-

161

tinue to show no indication of imminent attack. Yet an attack is about to commence just the same.

A rapid reaction task force centered around the Nimitz-class carrier *Theodore Roosevelt* now lies some two hundred nautical miles off the coast in the Arabian Sea.

Onboard are hundreds of battle-ready troops, already cammied up and good to go, as are hundreds more in transport ships attached to the strike package.

Already, a contingent of SEAL special forces hard-chargers have debarked the carrier, making their way toward the beachhead in camouflaged landing craft. Among these craft is a unique surface vessel called the Sea Shadow, seeing its combat debut, its angular black hull giving it a look of demonic malice, its engines virtually soundless.

In the air, an advanced AWACS E-3E aircraft has been orbiting the operations zone, receiving telemetry from other aircraft that ply the airspace, including forward deployed B-2B and F-117B stealth aircraft, Prowler and Hawkeye electronic intelligence planes, and the Global Hawk unmanned surveillance drone.

As AWACS receives telemetry from these airborne assets, its computers process and fuse the sensor data, presenting a real-time tactical picture to the *Roosevelt*'s commanding officer below.

In the darkened command and control centers of friendly forces at sea, the tactical picture is clear. But it is not so to enemy forces based on land. Prowler has used electronic countermeasures to lull unfriendly radars into a false sense of security and open a corridor of infiltration for the SEALs.

0410 hours.

Just before dawn the beachzone is lit by stuttering

flashes of automatic fire. The battle is brief and one-sided, resulting in heavy losses to the Iranian soldiers deployed on the beach and relatively minor casualties for the invading SEAL teams. Other special forces detachments have swept in via high-altitude, high-opening (HAHO) paradrop, gliding to their landing zones using MC-5 ram-air parachute systems. These forces have now secured the enemy command center inland.

Events now gather increased momentum, moving swiftly on multiple levels. Miles to the north, at the port city of Shabaz, F-117B Nighthawks destroy major communications nodes with precision laser-guided bomb strikes.

From the sea, the attack submarine *Helena* launches an advanced PAC-5 Tomahawk cruise missile with extended range and striking power. Other Tomahawks are launched from the *Roosevelt* and from the Aegis cruiser *Benfold*, stationed nearby.

As a high-flying SR-71 spyplane's real-time synthetic aperture radar imagery of the command center is relayed to the *Roosevelt*'s combat information center (CIC), analysts study the feed on a wide-screen, high-resolution computer monitor.

A structure in the foreground immediately captures their attention. Enhancing the digitized imagery, the analysts note the presence of millimeter wave dish antennas. Although the enemy has tried to conceal the antennas, it is clear that this structure houses a backup tactical communications node.

Two of the carrier's already-launched Tomahawk missiles are rerouted in midflight and remotely piloted to their destinations by naval Wizzos seated before computer consoles. The SR's next overflight will show a massive hole in the target building's flame-scorched roof.

As dawn breaks, the Iranian Navy (IRIN) recovers from its paralysis and locates the U.S. carrier task force. It immediately launches a counterstrike by sea and by air. Iran must act quickly. Already reports of troop landings along its coastline are flooding in via the country's severely damaged military communications network.

IRIN's plan of attack is based on the Russian model of saturation assault. Waves of MiG-30 and Sukhoi-35 fighter planes equipped with fire-and-forget weapons take off from hidden desert runways and vector out to sea while combat vessels on patrol are rerouted to close with the enemy.

In minutes, display screens in the *Roosevelt*'s combat information center identify the first hostile contacts, and the carrier's fighter crews prepare to intercept them.

F/A-23As are flying combat air patrol (CAP) at stations on the outer envelope of the carrier group's zone of protection. These fast, stealthy naval fighter planes are variants of the Joint Strike Fighter aircraft family.

AWACS-based flight controllers continuously track the contacts and the F/A-23As break formation to engage the enemy fighters, their pilots switching to real-time satellite telemetry for a clearer tactical picture simply by speaking directly to the planes' computers.

A digital window is instantly superimposed across a corner of the pilots' view of the section of sky directly ahead as seen through a lightweight head mounted display (HMD). Computer-generated tactical information is overlayed on the cockpit view in the form of real-time graphics and live video imagery.

The fighter pilots see Iranian MiGs out at 200

miles, the distance indicated by glowing green numerals below the crisp synthetic aperture radar (SAR) images of the bogies in the HMD window. Using eye movement to select AMRAAM long-range missiles from menu selections seeming to float before them in midair, the fighter jocks pickle-off the air-to-air munitions at standoff ranges.

Within minutes most of the Iranian fighters are killed before their pilots can even see the stealthy F/A-23As on their radar scopes. The remainder of the SUs and MiGs scatter and are taken out piecemeal by close-in AIM-9M missiles and computer-calibrated automatic weapons fire from the U.S. jets' boresighted nose cannons.

As the air battle progresses, the sea battle is joined. Fast Iranian cruisers close with the outer layer of ships screening the carrier task force while two Kilo-class submarines, the *Ormuzd* and the *Ahriman-2*, stealthily approach at a depth of 200 feet.

Canny enough not to get in too close, the commander of the IRIN battle group in the lead cruiser fires a salvo of Chinese-made Silkworm-3 antiship missiles as soon as the USS *Nevin,* one of the task force's Aegis vessels, is in range.

The missiles flare from the enemy cruiser's decks and skim the sea in a dodging, leaping, radar-evading flight path intended to protect them from the rapid defensive fire of deck-mounted Phalanx CIWS (close-in weapon system) guns. In seconds, dozens of smart weapons are locked onto the ships of the U.S. flotilla and are speeding toward their targets.

0502 hours.

In the combat information center (CIC) of the Aegis cruiser *Benfold,* tactical display screens at stations all around the darkened war room light up with hostile contacts. The CIC's large, flat-panel ''god'' screen shows the combined tactical picture of the battlecube, displaying all the contacts seen by the Benfold's next-generation SPY-2 radar.

The *Aegis*'s advanced radar system can track more than 200 simultaneous contacts, and there are more than fifty now being plotted. As a result, the defenses of the carrier task force are mobilized. Rockets are fired from the *Benfold*'s decks while Phalanx guns shoot incandescent streams of bullets at the sea-skimming Iranian missiles.

Below the surface, the *Helena* has launched a Mark 3 torpedo at one of the Kilo-class submarines. The sonar-guided and acoustically homing undersea munition strikes the *Ormuzd* amidships as it fires its own torpedo, then turns and tries to run. The Mark 3 explodes on contact, rupturing the Kilo's hull and slicing the sub in half.

As the *Ormuzd* sinks to the bottom, the torpedo she's launched at the *Helena* loses target tracking and acquisition. Confused by decoys that the *Helena* has ejected, the torpedo plunges toward the seabed and detonates harmlessly in the depths below.

The surviving Kilo has meanwhile maneuvered for cover, but a Manta remote-operated vehicle also deployed by the *Helena* has found it, maneuvered close to its hull, and launched a kinetic energy (KE) weapon, triggering a secondary explosion in the boat's torpedo room. The one-two punch of twin explosions instantly turns the second Kilo into a spiraling cloud of underwater debris.

0535 hours.
The entire series of multiple engagements on sea,

in the air, and on land has lasted only a brief while, but it has left the opposition forces in a state of nearly total disarray. Coalition troops, mobile armor and combat materiél of all kinds are now piling up on the secure beachhead and armored spearpoints are thrusting their way inland at the speed of a Sunday drive.

At the White House, President Al Gore scans the text of a televised address as he prepares to go on camera. Only minutes before, he has learned of a terrorist poison gas attack on a large U.S. city that was stopped cold by counterrorist assets of the FBI and local police acting on information supplied by the CIA. The danger of fresh terrorist offensives as the war in Iran continues cannot be ignored, but the president is confident that these, too, can be prevented by vigilant security forces.

He will truthfully report to the nation and the world what the Iranians have just learned the hard way— America is good to go when its interests and those of its allies are threatened by the hostile actions of rogue states.

The Ooda Loop:
Forward Deployment in the Battlecube

The scenario above was taken from Fleet Battle Experiments (FBE) conducted by the USN. The FBEs are being conducted to develop new concepts and integrate high-technology advances under the Navy's Copernicus architecture.

The specifics of engagement as described above, or the weapons systems therein, might or might not be actual elements of tomorrow's MRC, but the framework of operations is a realistic projection of events as naval FBEs have foreseen them.

The scenario illustrates how joint operations in the multidimensional battlecube enable carrier-based expeditionary forces to leverage high-technology assets in order to prevail against an alert but technologically less advanced adversary. It is precisely such an adversary that the U.S. and her Coalition allies are likely to be confronted with in future regional theater conflicts.

The events that precipitated the attack were not mentioned, although any number of possibilities for a crisis situation might have supervened.

For example, Iran might have acted to cut off the flow of its crude oil from inland pumping facilities to ports located along its coastline. Or that country might have threatened to use its growing stockpile of nuclear and even chemical and biological weapons against friendly states, such as Saudi Arabia, Jordan, Egypt, or Israel. Alternatively, surveillance data might have indicated that Iran was preparing to attack one of its Mideast neighbor states in a preemptive military strike.

Whatever the event or combination of events that triggered the decision to take the military option, U.S. forces were in no position to play games of wait-and-see. Nor did this particular chain of events give the U.S. and her allies sufficient time to put forces in place and build up supplies in a Desert Shield–style operation.

Here, a response would need to be made quickly and credibly to stop the opposition cold before it could carry out a rash act with severe, negative global consequences. Forward positioning of the *Roosevelt* carrier task force was key to securing this objective.

Because it was prepositioned in the Indian Ocean, the *Roosevelt* and her escort vessels could steam to the Arabian Ocean in sufficient time to meet the crisis in its developmental phase.

Once in the combat theater, expeditionary forces began phasing through the OODA loop—observe, orient, detect, act.

JMCIS—the joint maritime combat information system—integrated with Copernicus command and control architectures, rapidly brought real-time, multiplatform, sensor-to-shooter data to all friendly combatants. High-altitude spy planes, unmanned aerial vehicles, fighter planes, electronic warfare aircraft, and airborne warning and control planes swept through various levels of the tactical envelope, their sensors unfazed by the darkness.

Data from these multiple surveillance platforms were collected by AWACS, Hawkeye, and the computer systems located onboard the *Roosevelt* and *Benfold*. This was then fused into a comprehensive tactical picture of the area of operations using JMCIS and made available to battle managers on computer screens and radar displays.

At the same time, information warfare jammed and confused unfriendly radar with invisible electronic attack, blinding and deafening the enemy to the presence of the carrier group as it approached from the sea. Command, control, and communications (C3) nodes, radar emitters, power grids, and tactical computer systems all became targets in a multipronged assault across the electromagnetic spectrum.

Once the assault force had gained information dominance, the strike commenced. Seaborne special operations forces infiltrated Iranian territory under cover of darkness in stealthy assault craft. Once the commando forces had secured their objectives, the main strike elements attacked in earnest. Stealth aircraft used the hole punched through Iranian radar coverage to enter hostile airspace and then deal the enemy's main warfighting infrastructure a devastating blow.

Block 4 TLAMs, advanced variants of the Tomahawk land attack missile, cruise missiles which could be configured while in flight from seaborne tactical command centers—were launched on terrain-following paths to their targets farther inland.

While in flight, some of these brilliant air-breathing robot munitions were rerouted from the carrier *Roosevelt* to deliver a knockout blow to enemy command nodes already severely damaged by pinpoint-accurate hits from laser-guided bombs dropped by B-2 Spirit and next generation F-117B Nighthawk stealth planes.

Finally, when a disoriented and sensory-deprived enemy succeeded in marshalling its severely damaged forces to launch a brute-force counterstrike in the cumbersome Soviet mode, attempting to saturate the attacking forces with missiles and so overwhelm their defenses, this strategy proved no match for agile surface combat systems that could provide a defense in depth for the carrier force.

With the enemy's counterattack neutralized, amphibious elements and paratroops were rapidly shuttled into the theater and an advance deep into unfriendly territory commenced virtually unopposed.

The Carrier Battlegroup
in Attack and Defense

The carrier battlegroup that has played such an important role in assuring U.S. dominance in recent global conflict had its origins in the Second World War. Carriers have played major roles in every international conflict that has arisen since that time, from action in the Tonkin Gulf during the Vietnam years to the war in the Persian Gulf and in the several peacekeeping operations that have followed in its wake.

The carrier will be with us for some time to come. In fact, tomorrow's major regional conflict will see the carrier play an even greater role in military operations across the globe.

New weapons, such as improved Block 3 TLAMs (Tomahawk land attack missile), will give it a more formidable force projection capability than ever. New C4I systems tied into the JMCIS system will give it mastery of the battlecube. New aircraft will rise from its flight deck and scream into the night to drop bombloads of advanced laser-guided ordnance on distant targets or to engage hostile planes at long range.

These aircraft will be fast, stealthy, and maneuverable naval variants of the Joint Strike Fighter. Some versions will feature STOVL (short takeoff/vertical landing) capability and lift straight off the carrier's decks.

Accompanying the carrier, screening her from harm, will be advanced cruisers and destroyers of the Arleigh Burke and Spruance class whose SPY-2 and higher radars will enable them to track multiple targets while attack submarines ply the undersea channels.

New too will be arsenal ships, floating supply depots carrying stores of weapons, equipment, food, vehicles, and every form of materiél needed to wage war effectively. Tomorrow's carrier group will be a seagoing fortress, able to conduct operations on the sea, on the land, in the air, and under the water.

The Carriers

In the future, as today, the heart of the carrier battlegroup will probably be one of the seven Enterprise-, Kennedy-, Kitty Hawk-, Forrestal-, or Nimitz-class carriers now in service.

It is unlikely, given the recent defense-spending cuts mandated by the QDR, that the two mew Nimitz-class carriers now being laid down will reach service in time to participate in the next MRC. Those carriers that have played major roles since the Vietnam era will probably be with us as the twenty-first century dawns.

Measuring 1,040 feet in length and displacing 95,000 metric tons of water, Nimitz-class carriers are by far the largest warships in the world. The area of their flight decks is the equivalent of four and a half acres, with enough capacity for about seventy aircraft of varying types. Belowdecks, there's enough space for the more than 6,000 crew members generally found onboard in wartime.

The Nimitz ships are fast, too, capable of attaining speeds of some 30 knots, the equivalent of approximately 35 miles per hour.

Powered by twin nuclear reactors, they can remain at sea for long periods of time and need to be refueled only every four years. The reactors power eight steam turbine generators, each producing 8,000 kilowatts of electrical power—about enough electricity to sustain a small city.

Defensive weapons systems include radar-guided, short-to-medium-range Sea Sparrow missiles and Phalanx close-in weapons system (CIWS) guns for near-at-hand antiaircraft and antimissile defense. Each Phalanx "R2D2" mount contains integral search and track radar and a six-barrel 20 mm Gatling gun firing 3,000 rounds per minute (which translates to fifty bullets per second).

The air wing attached to the Nimitz carriers will vary, but a typical configuration will consist of nine squadrons comprising F/A-18 Hornet, F-14 Tomcat, EA-6B Prowler, S-3 Viking, E-2 Hawkeye, and SH-60 Seahawk helos.

A system of steam catapults is used to launch strike fighter aircraft from the carrier's deck. The catapults give the fighters a short-takeoff capability they would not otherwise possess. On landing, motor-powered arresting cables that tighten when placed in contact with the warplanes' landing gear are used to safely recover the arriving jets.

During military operations, fighters will fly CAP or combat air patrol to warn and protect the carrier group against unfriendly planes. They will also conduct long and short-range strikes, support land forces, lay sea mines, and engage in antisubmarine operations.

Hawkeye aircraft will provide the carrier group with aerial surveillance and Prowler planes will engage in electronic warfare while Seahawk helicopters and Viking antisubmarine planes will drag dipping sonar modules by long cables to scan below the surface of the sea for prowling unfriendly subs.

A carrier battlegroup with a nuclear carrier at its center covers an extremely vast stretch of open ocean. The battlegroup spreads out over an area with an approximate two-hundred-mile radius. This area represents the approximate distance from Richmond, Virginia to Washington, D.C. and back again.

But this coverage isn't only two dimensional. It's a layered defense structure in three dimensions: on the surface, in the air, and below the sea, stretching vertically to 70,000 feet and below to the maximum operating depth of the submarines guarding the battlegroup.

At the top of the envelope, combat aircraft fly CAP. The all-weather F-14 Tomcats and F/A-18 Hornets fly their stations in a circular pattern, each keeping its nose to the other's so that their long-range threat-identification radars are always providing a full 360-degree coverage.

There will generally be a minimum of two CAP stations flown at any particular point in time, at various parts of the envelope. Hawkeyes also occupy the outer zones of the umbrella of protection surrounding the carrier, with Viking subhunting aircraft stationed nearby.

On the sea below, in the middle range of the coverage umbrella, fast Ticonderoga-class Aegis battle cruisers equipped with advanced SPY radars and armed with missiles watch for incoming missiles, planes, and seagoing warcraft.

Named for the shield carried by ancient Greek warriors, Aegis cruisers will be among those new main ships of the line scheduled for building into the year 2010. The guided missile cruisers are multimission surface combatants capable of supporting carrier battlegroups or amphibious forces, or operating independently as flagships of surface action groups. The cruisers are equipped with Tomahawk TLAMs for strikes inland from seaborne launch sites.

Close to the heart of the envelope, anything that succeeds in penetrating in deep will come into conflict with the carrier's final line of defense. These are guided-missile frigates, which are single-mission platforms outfitted for air warfare roles.

The least versatile of all those combat ships tasked

with screening the carrier from harm, they are easily the toughest, as illustrated when the USS *Stark* survived two direct hits by Iraqi Exocet cruise missiles during an action in the so-called "Tanker War" in the Persian Gulf that took place between 1984 and 1988.

Destroyers patrol in this segment of the envelope, too. These fast warships are optimized for antisubmarine warfare, or, when equipped with guided missiles, for antiship and antiair combat. Those, such as the Spruance class, equipped with the MK-41 vertical launch system, can fire Tomahawk cruise missiles to strike targets far inland. The most advanced types, the Arleigh Burke class, are equipped with the Aegis combat system and the SPY-1D multifunction phased array radar.

When unfriendly contacts are sighted, the carrier group takes immediate action. Cruisers from a picket line along the threat axis—the line facing the direction of incoming fire—as the carrier moves to the rear of the group and additional surveillance aircraft and warplanes take off into the skies.

Below the ocean's surface, the battlegroup's submarine escort watches for hostile subs while Seahawks dip their sonar probes and P3-Orions scan for hostile undersea vessels. If the enemy force launches a strike or does not move off when warned away, the carrier group's defenses are then activated. Missiles, aircraft, ships, and other combat assets are all mobilized to the defense of the battlegroup.

Threats

As formidable as it most certainly is, the carrier battlegroup has numerous inherent vulnerabilities.

During the previously cited Tanker War in the Persian Gulf, precipitating U.S. involvement by Iran's refusing Western tankers access to the oil facilities along its coast, the United States protectively reflagged foreign tankers as American ships to place them legally under her protection, then the USN escorted them through the waterway to the open sea.

During this action, an Iraqi Mirage F-1EQ fighter plane launched Exocet missiles at the USS *Stark,* a frigate attached to the U.S. carrier group in the area, and scored two direct hits on the warship. Later on, another frigate, the *Samuel B. Roberts*, struck a mine while participating in the same operation. In the wake of these events, the commander of the Aegis cruiser *Vincennes* authorized a missile attack on an unarmed Iranian commercial jet after one of the cruiser's helicopters was attacked by Iranian gunboats. The jet had been misidentified by the *Vincennes*'s SPY-1 radar system as a warplane, and the ship's commander expected imminent attack.

For a time these incidents raised questions about the validity of large ships being deployed in an era of smart, low-cost ship killers like Exocet, and also about the danger posed to civilian targets by automated combat systems. While countermeasures have improved since the days of the tanker war, these questions about survivability and lethality still linger. The threats have also gotten smarter, faster, and cheaper and will continue trending in this direction out into the future. The widespread availability of electronic technology to even second-rate powers will ensure this development.

Mines are one of the best examples of this danger. Of the forty-nine countries known to possess mining capabilities, at least thirty are mine-producers, a number that has continued to grow since the Gulf War. Also contributing to the proliferation of mines is the great Russian arms selloff in the wake of the Soviet Union's breakup. The selloff has included portions of a vast stockpile of some half million sea mines ranging from moored contact mines of World War One vintage to the most advanced types found today.

Mines are among the deadliest threats to the carrier battlegroup, and these are growing deadlier all the time; in fact, it has been estimated that some 75 percent of damage to U.S. Navy capital ships since 1988 came from mine strikes (including one from a $1,500 World War One version that caused approximately $96 million in damage to the USS *Samuel B. Roberts* in a Tanker War action of 1988). The more sophisticated mines produced today are even more dangerous than sea-skimming missiles like Exocet. They can lurk hidden on or near the sea floor, probing the electromagnetic and acoustic spectrums for information on what type of vessel is passing overhead until the correct preprogrammed sonar or audial signature triggers an explosion.

Today's mines are nothing like the "dumb" mines of the first and second world war eras. They are sophisticated robotic predators that lie in wait like submerged mechanical spiders until their prey comes near.

Equipped with fast computer processors linked to magnetic, LIDAR (light detection and ranging), audio, and synthetic aperture sonar sensors, these mines are capable of distinguishing the specific types of vessels they are programmed to destroy from those in which they are not interested.

An antiship mine such as the Captor type will typically be found floating fifty feet above its mooring line, which is anchored to the sea bottom by steel holdfasts. Its cylindrical housing may be filled with high explosive, or it may alternatively contain a torpedo that can be fired when its target sensor and engagement system acquires the proper type of ship passing overhead. At that point the mine either explodes or launches its torpedo at the target vessel.

Mine countermeasures come in a variety of types, including minesweeping vessels with glass-composite hulls designed to reduce their hull signature to magnetic sensing mines below. The USN and other navies recognize that countering the threat of tomorrow's mines, which will be even harder to kill than today's, is a high priority in maintaining readiness to counter emerging threats.

The USN is examining the use of Superconducting Quantum Interference Devices (SQUID) which are towed behind ships to jam mine sensors and neutralize the mines.

The SWEEPOP system is geared first to detonate mines detected in a minefield using a Mine Actuation Registration System (MARS) and then create a database of the types and explosive charges identified so that similar mines can later be pinpointed and destroyed.

Submarines present an even deadlier threat to the battlegroup than do mines and missiles combined. The former Soviets lead the world in the design and production of silent electric-powered boats that even the smallest rogue states can afford.

The top of the Russian line is the Kilo-class diesel-electric submarine developed by the Rubin design bureau. The subs are pitched to potential buyers by Rubin sales representatives at international arms bazaars such as the Satory '97 arms exhibition.

Representatives of the Russian underwater technologies bureau Morphyspribor and torpedo manufacturer Gidropribor have stated that the newest Project 636 Kilos embody sweeping upgrades to approximately forty critical subsystems.

Ten Project 636 Kilos are either in service or on order today, although with specifically which national purchasers has not been revealed. Since the U.S., France, Britain, and other Coalition countries don't officially field the Kilo, it can be inferred that unfriendly states are among those with whom the Russians are currently conducting business.

Being quiet is one of the Kilo 636's deadliest virtues. Its double-hull construction provides the boat with high reserve buoyancy. This means that in emergencies it can surface with even a single one of its six watertight compartments unflooded.

Kilo's teardrop-shaped hull makes it more maneuverable than other subs in its class and greatly reduces its noise signature, with anechoic rubber tiles applied in critical areas to reduce the signature still further. The sub also has a specially designed seven-bladed propeller that enables it to move as easily through shallow channels as it can through open ocean.

The propeller incorporates technology that stabilizes its rotational and vibrational levels and further reduces noise emanating from the sub. According to technical specifications, the 636 is so quiet that when running at six knots it is at least eight decibels more silent than any preceding electric boat produced by the Soviets.

The Kilo is also optimized for antiship warfare. Its computerized sonars can counterdetect unfriendly subs at five times the ranges of the German, French,

American, or other Western subs that are its closest competitors. Rubin makes the claim that the 636 can see even a stealthy low-noise-signature submarine at underwater distances approaching thirteen miles. It can then strike at its target using a suite of deadly weapons including five kinds of mission-specific torpedos.

The U.S. and her Coalition partners are paying close attention to the innovations in electric undersea boat technology represented by current Russian development efforts like Kilo. The New Attack Submarine, which will succeed the U.S. Seawolf attack submarine, will be equipped with technology devised to counter the threat posed by the advanced Kilo-class subs.

On Submarine Warfare

Strategic deterrence has been the primary mission of the fleet nuclear ballistic missile submarine (SSBN) since its inception in 1960. The SSBN provides the nation's most survivable and enduring nuclear strike capability.

The Ohio class SSBN replaced aging fleet ballistic missile submarines built in the 1960s. The Ohios are far more capable boats than their predecessors in every regard. They are wider ranging, are capable of staying submerged for longer periods, make far less noise, and are better armed. Equipped with the Trident ballistic missile, they can strike strategic targets thousands of miles away.

Ohio-class/Trident ballistic missile submarines provide the sea-based "leg" of the triad of U.S. strategic offensive forces. By the year 2000, the eighteen Trident SSBNs currently in service (each carrying twenty-four missiles) will carry 50 percent of the total U.S. strategic warheads in service.

Although the missiles have no preset targets when the submarine goes on patrol, the SSBNs are capable of rapidly targeting their missiles should the need arise, using secure and constant at-sea communications links.

The first eight Ohio class submarines were originally equipped with twenty-four Trident IC ballistic missiles. Beginning with the ninth Trident submarine, the USS *Tennessee* (SSBN 734), all new boats are being outfitted with the more advanced Trident IID-5 missile system as they are built. Earlier Ohio boats are being retrofitted to Trident II–series missiles. Trident II can deliver significantly more payload at more extensive ranges than the Trident IC, and do this with greater accuracy.

The Ohio-class submarines are specifically designed for extended deterrent patrols. These huge undersea vessels must essentially function as seaborne military bases, providing their crews with most of the services and facilities to be found at a land-based installation.

To decrease the time spent in port for crew turnover, maintenance and replenishment, three large logistics hatches are fitted into the Ohios' hulls to provide large-diameter resupply and repair openings. These hatches allow sailors rapidly to transfer supply pallets, equipment replacement modules, spare parts for machinery, and other stores, significantly reducing the time required for replenishment and maintenance while docked.

The design and systems architecture of the Ohio-class subs and advanced hull construction materials and techniques afford the nuclear submarines with a mean time between overhauls of approximately fifteen years.

The concept of technical superiority over numerical superiority was and still is the driving force in American submarine development. A number of Third World countries are acquiring modern state-of-the-art non-nuclear submarines like the Russian Kilo. Countering this threat is the primary mission of U.S. nuclear attack submarines. Their other missions range from intelligence collection and special forces delivery to antiship and first-strike warfare.

The USN began construction of Seawolf class attack submarines in 1989. Seawolf is designed to be exceptionally quiet, fast, well-armed, and equipped with advanced sensors. It is a multimission vessel, capable of deploying to forward ocean areas to search out and destroy enemy submarines and surface ships and to fire missiles in support of other forces in theater. The first of the class, Seawolf (SSN 21), completed its initial sea trials in July 1996.

Attack submarines also carry the Tomahawk TLAM cruise missile. Tomahawk launches from attack submarines were successfully conducted during Operation Desert Storm.

Of special interest are the boats of the Benjamin Franklin submarine class, which were converted from older fleet ballistic missile submarines and which carry drydock shelters for other subs. These boats are also equipped for special operations and are specially fitted to carry the swimmer delivery vehicles (SDV) used by SEAL teams for covert sea-to-land infiltration. The former missile spaces of Franklin-class subs have been converted to serve as accommodations, storage facilities, and recreation areas.

Toward the Future and Beyond

Tomorrow's seaborne operations will be more complex than ever before in naval history. They will not only encompass historical and traditional actions against other surface and undersea vessels, but will play increasingly large roles in strategic lift of expeditionary forces, support of airborne and land-based assault elements, and direct offensive operations through missile strikes on distant inland targets.

The USN's concept of "From the sea" will encapsulate the essence of America's ability to project its military power globally, where and whenever the need arises.

But in tomorrow's combat operations, the threat environment can transform itself from moment to moment. No development will be predictable with any degree of absolute certainty, and all forces in theater will need to assume many roles. The future of war is chameleonic, changing its colors and shapes with increasing rapidity. Before very long, tomorrow's war will have already changed into war beyond tomorrow.

In our final chapter, we will try to peer somewhat deeper into this future landscape of war and divine what it may hold for tomorrow's soldier, wherever he may fight.

EIGHT

•••••••••

War Beyond Tomorrow

ACCORDING TO MOST current military thinking, future war will be both regional and conventional in nature. It will be fought and won quickly using the overwhelming superiority of American and allied weapons systems to prevail against regional threats.

This weaponry will be conventional, that is, it will be non-nuclear. Its lethality will be enhanced by combining chemical explosives with computerized targeting and delivery systems and by the use of stealth to deny the enemy the means to defend its warfighting infrastructure or to effectively counterattack.

Such thinking is a kind of military perestroika, a new thinking in the wake of the revolution in military affairs or RMA that began to take shape after the breakup of the Soviet Union changed the bipolar superpower relationships of the Cold War into the multipolar regional relationships of the New World Order. This thinking solidified into doctrine in the wake of the Gulf War. But prior to these revolutionary developments, a completely different set of suppositions about future war prevailed.

Throughout the 1980s, during the Reagan years, the theories of the nuclear strategists that forged strategic policy for the preceding Carter Administration had developed into the doctrine of winnable nuclear warfare. A rough-and-tumble definition of what "winnable" meant would be for our side to survive a nuclear exchange and inflict sufficient damage to the other side—that is, the Russians, to force them to sue for peace. An acknowledgment that a confrontation between the superpowers was imminent and that it would quickly escalate beyond the conventional threshold into a nuclear exchange drove U.S. war-planning and weapons procurement efforts.

Many of the weapons systems that current war-planning sees as winning tomorrow's major regional conflicts owe their inceptions to use in nuclear warfighting. Stealth aircraft, for example, were developed to assure penetration of Soviet ground-to-air missile defenses and deliver nuclear weapons to priority targets. But there were other schemes and systems that are no longer on the table.

Land-based ballistic nuclear missiles are no longer being developed by the United States or Russia as a result of disarmament treaties signed by both nations just prior to the Gulf War. This is truly a revolutionary development in military affairs, as for decades since the end of the second world war these missiles had been an obsession of both superpowers, culminating in the development of sophisticated mobile launch systems by both sides.

As Gulf War experience with mobile Scud launchers showed, destroying mobile missile systems is an extremely hard thing to accomplish with a high degree of certainty. And this was on the theater level and against launch systems far less sophisticated than those formerly in development for nuclear weapons.

The nuclear warfighting scenarios developed during

the Carter and Reagan years would have meant attacking mobile launch systems on the *strategic* level, that is, at global distances. Considering the sophistication of the rail-mobile SS-25 missiles then being fielded by the Soviets and the MX system being developed by the United States, the concept of destroying enough missiles to make a difference was nothing short of a deranged fantasy on both sides.

Actually, neither side was lunatic enough to believe that "winning" a nuclear war through countervailing strikes would work, which was why the Reagan Administration began the development of an entirely new concept in ballistic missile defense, one that—by means of a Rube Goldbergesque array of laser, nuclear, microwave, kinetic energy, and other weapons that would destroy incoming missiles at the apogees of their boost phases—would act as a space-based shield or umbrella against nuclear attack.

This was the Strategic Defense Initiative, called SDI by the strategists and Star Wars by practically everyone else. Part bargaining chip, part special effects extravaganza, and part reality, SDI was disparaged by critics from its inception as being too complex to be workable and too farfetched to be possible using existing technology.

Yet SDI has been cited as one of the factors in pushing the Soviets to the nuclear bargaining table by confronting them with a show of technological superiority they could not hope to match. Clearly, the U.S.S.R. *did* believe that SDI was more substantial than critics in the United States and Europe believed at the time—and it should be noted that the Soviets have had an excellent track record with regard to knowing about the most closely guarded U.S. military secrets, the Manhattan Project to develop the first atomic bomb being just one example.

Strategic nuclear attack and space-based defensive shields are currently out of fashion in military vision statements—*officially* out of fashion, that is. They are no longer heard much about in Pentagon white papers concerning current and future warplanning.

Politically, debate on weapons of mass destruction (WMD) is suicidal to careers and lethal to procurement budgets. The electorates of the Western democracies want to hear about quick, winnable conventional wars in the Gulf War mode. After nearly a century of the most devastating warfare since perhaps the religious wars of the sixteenth century or the Viking incursions of the eighth century, global populations are sick of "Big War" and have tuned it out of their collective consciousness.

Unofficially, however, nuclear weapons and the militarization of space are much on the minds of warplanners. They have to be, because more than enough nuclear weapons currently exist to reduce Moscow and Washington to radioactive embers, and because some of these weapons are almost certainly now in the hands of rogue states like Iraq and Iran and also potentially in the hands of transnational terrorist groups.

There is also a strong likelihood that nuclear weapons will be tactically deployed if tomorrow's MRC takes certain turns for the worse. Bunker-busting nukes that detonate below ground and leave behind relatively low residual radiation levels are already waiting in America's arsenals. These are called HDBTDC weapons because of their "hard and deep buried target defeat capability." Other nuclear devices called SADMs are small enough to be part of tomorrow's soldier's personal weapons load and getting smaller all the time as secret programs shrink them down.

In addition to the nukes, there is also a nasty array of chemical and biological weapons in the arsenals of friends and potential foes alike. Despite attempts to ban such weapons, proliferation on overt and covert levels continues. During the Gulf War the use of these weapons by Iraq was especially feared by the allies, since Saddam had previously demonstrated the will to use them in Iraq's war with Iran and against Kurdish tribesmen in Iraq.

Space-based weapons and the militarization of orbital and suborbital space are also certainties, given the increasing vulnerability of military communications and intelligence satellites to preemptive attack. As of this writing, the U.S. Department of Defense has announced (albeit elliptically) what has been rumored for years; that it has the nascent capability to destroy missiles and satellites using military lasers mounted on high-flying planes.

Other nations also have, or shortly will have, satellite-killing missiles of their own, while it is an open secret that the Russians have had an array of satellite-killing satellites in orbit for at least a decade. Given these facts, and the need to maintain global surveillance in times of peace and war, space militarization will be part of tomorrow's warfighting imperatives.

Beyond this, the increasing reliance on computerized systems for command, control, communications, and intelligence will increasingly make military information processing systems the target of what has been termed "information attack." Civilian computer networks on which electronic media and global commerce depend will also be chosen as targets by national and subnational or terrorist military arms. At the same time, friendly information attack capabilities will be perfected for use against our enemies' computer networks.

This form of warfare, called "information warfare" (IW), has been mentioned elsewhere in this book as a component of the digital battlefield. But IW has the potential to transcend this relatively narrow application. It could unleash a chain reaction of information attack on the world as deadly as some nuclear scenarios might prove, insofar as it can cripple the means by which the basics of modern life are provided to the world's civilized populations.

War beyond tomorrow will be a blend of the old nemeses and the new, combined in terrifying ways that will pose daunting challenges to soldiers, who will be everyone, on every part of the battlefield, which will be everywhere.

NBC Warfare

NBC warfare is nuclear-biological-chemical warfare. NBC weapons can be based on one, two, or a mix of all three principles. To rogue states with limited military budgets and little interest in the more humanitarian approach to war favored by the West, it's an attractive alternative to stealth and other precision strike weapons, which they cannot afford.

NBC is inaccurate and indiscriminate in who it kills, but it gets the job done. For this reason it's also a weapon of choice for terrorists, such as the Japanese Aum Shin Rikyo (Supreme Truth), which on March 20, 1995, staged a lethal chemical attack on the Tokyo subway system.

During the Gulf War, a declared policy of massive retaliation (meaning a nuclear strike in political code) if Saddam used the chemical weapons known to be in his arsenal against Coalition members or the Israelis was theorized by some to have kept the warheads on Iraqi Scud missiles conventional. But the next confrontation with another would-be conqueror might not meet with even such a limited degree of rationality. For this reason and others, it's realistic to assume that tomorrow's MRC is likely to see some form of NBC warfare used.

Chemical And Biological Agents

The extreme deadliness and toxicity of even the most common and least exotic chemical agents can be judged by the mysterious illness called Gulf War Syndrome that has afflicted many Desert Storm veterans. It's now believed that more than 15,000 U.S. troops in the Gulf were exposed to trace amounts of the nerve agent tabun and the blister agent mustard gas after Iraqi weapons dumps were hit by air strikes during the conflict.

Czech chemical warfare detection teams had been patrolling the northern Saudi desert throughout the war. Their mission was to act as forward observation units to provide advance warning of imminent chemical attack. The Czechs were then, and are now, rated among the most expert in chemical warfare capabilities in the world, and their detection equipment is considered highly reliable.

In the early days of the war, during the Desert Wind air strikes on Iraqi military targets, the Czech NBC teams flashed warnings to Coalition forces under the

unified command of General H. Norman Schwarzkopf. On those occasions their instruments had detected small quantities of nerve and mustard gas
wafting across the border from Iraq into the vicinity
of American troops.

Nevertheless, their warnings were ignored by Central Command, the Coalition military oversight body.
Even as the Czechs rushed to put on their MOPP
gear—gas masks and chemical warfare suits—after
detecting the toxic gas seepage, U.S. troops stationed
nearby remained unprotected and unconcerned. They
had not been ordered to suit up.

In the years following the war, hundreds of veterans
began to complain of medical symptoms of various
kinds. At first such claims were dismissed as unrelated
to anything that had happened in the Gulf. But in the
wake of class action lawsuits and extensive press coverage, the Pentagon has recently given such claims
credence and will issue monetary compensation based
on Gulf veterans' charges of exposure to chemical
warfare agents.

If such small quantities of nerve agents could cause
such lasting symptoms (possibly heightened by reactions with pyridostigimine bromide, a partially tested
nerve agent vaccine with which some 400,000 Gulf
War servicemen and -women were injected by U.S.
medical teams), actual combat exposure to some of
the even deadlier agents in the world's chemical arsenal could be inconceivably devastating. It's estimated that by the close of the Cold War, as much as
30 percent of the available tactical missile warheads
in the world's arsenals were chemical, and most of
these weapons stockpiles have never been destroyed.

Tomorrow's enemies realize the harm chemical attack can do and are ready to take advantage of it. In fact, a few states have already put NBC weapons to the test. Chemical weapons were used in 1967 by Egyptians in Yemen, in 1979 by Vietnamese in Laos, in the 1980s by Soviets in Afghanistan, and in 1990 by Iraq against the Kurds; the British also used poison gas against Kurdish tribesmen in Iraq after the end of World War I, and even the United States has been accused by the Vietnamese of using CS (tear) gas in concentrations high enough to prove lethal. More recently, the Japanese terror army, Aum Shin Rikyo used a home-brewed batch of the deadly chemical agent tabun in its attack on the Tokyo subway system.

Furthermore, despite initiatives such as the U.S.-backed 1993 Chemical Weapon Convention (CWC), ratified by the U.S. Senate in April 1997, and the internationally sponsored 1972 Biological Weapons Convention, biochemical weapons research, testing, and stockpile-building continues by both rogue and "legitimate" states, the United States included.

Desert Storm has shown that such dice-throwing with the devil could well result in the opening of a nuclear Pandora's box. Iraq's known gassing of its Kurdish villages led the U.S. to tacitly (and unofficially) threaten retaliation with tactical nukes should Iraq deploy biochemical agents against U.S. troops. Israel sent similar signals by covert "backchannel," warning Saddam of what would happen if Iraqi Scuds were armed with biochemical warheads. Had Saddam not shown restraint, the door to nuclear war could have opened in the Middle East. The deadly genie, bottled up since Hiroshima and Nagasaki, would have finally been let out once again.

Know Your Poisons

Chemical agents are classed by physiological effects: lethal, choking, hallucinogenic, necrotic, and other disabling symptoms.

Volatile agents are easily vaporized liquid substances with viscosities compared to gasoline. They are dispersed so finely by the weapons that deliver them that they evaporate instantly. Distributed by wind throughout their target areas, they work through attack on the respiratory system. Their effects can last for minutes or for up to several hours.

Persistent agents have viscosities like that of motor oil. An exploding weapon sprays fine droplets of the agent over the target area, contaminating clothing, implements, and terrain. The effects of these agents can last for a period of weeks.

Biological warfare (BW) agents are generally derived from naturally occurring toxins, such as snake or fish neurotoxin, or viruses, such as bacteriophages. Neurotoxic or nerve agents come in two general forms, postsynaptic and presynaptic.

Postsynaptic neurotoxins, which can be derived from cobra and sea snake venom, work by disrupting nerve signals at synaptic junctions where nerve endings meet muscle tissue, blocking the neural receptors. Presynaptic neurotoxins work by first accelerating the neural activity at synaptic junctions, and then halting it. Their effects are similar to those of postsynaptic neurotoxins.

Muscle paralysis affecting all biological functions results from both forms of BW attack, and victims can literally drown in their own saliva. Necrotic effects, such as myonecrosis—severe muscle damage resulting in the loss of fingers, toes, and limbs—can also result, as can severe hemorrhage throughout all organ systems. After exposure, the human body literally becomes a bag filled with blood pouring from ruptured veins and suppurated organs.

On the viral level, bacteriophages are viruses that infect and multiply within human cells, disrupting cellular activity and killing the target by effects including severe muscular contractions, hallucinations, paralysis, and hemorrhaging. Viral agents are militarized, bioengineered versions of naturally occurring strains.

Policies and Programs for NBC Defense

Since more than twenty-five countries are believed to now possess, or to be in the process of developing, NBC weapons and the means to deliver them on the battlefield, and since these weapons are effective in countering U.S. and allied conventional superiority, the threat of NBC attack in tomorrow's war cannot be ignored. All force modernization efforts mandated by the U.S. Department of Defense (DOD) must now incorporate NBC survivability into equipment design to afford reasonable protection against the disarmament of U.S. troops under NBC attack.

Early detection and avoidance are the two cornerstones of U.S. NBC defense policy, and a number of DOD-DARPA research and development projects are currently underway in this sector. In developing CODA, the chemical/biological operational decision aid, DARPA is attempting to provide the Chemical Biological Incident Response Force (CBIRF)—a deployable force capable of performing chemical or biological consequence management following NBC attack—with a knowledge-based system to enable rapid response to all forms of NBC attack.

CODA's database incorporates detailed models of human response to symptoms of exposure, their severity, and their effects on performance and the degree of their lethality. The computerized system also estimates the length of time that CBIRF personnel can operate effectively in protective gear.

CODA estimates the downwind NBC hazard using standard predictions called "plume models" and generates casualty predictions for critical civilian groups and military personnel. The CODA database is constantly updated by orbiting satellites and other sensors to provide a real-time picture of the developing battlefield situation.

In the contamination avoidance sector, there have been a number of new technological concept demonstrations entering the final stages of development. One of the most important of these is Joint Service Lightweight Integrated Suit Technology (JSLIST), a program whose goal is to provide U.S. forces in all service branches with a common chemical protective ensemble, including suit, boots, gloves, and mask, that will allow the integration of chemical protective clothing as part of the standard duty uniform. This means that instead of wasting precious time climbing into bulky MOPP gear, tomorrow's soldier will already be wearing most of the NBC protection he needs.

In addition to prediction, assessment, and response to NBC attack, there are also projects under way to develop medical countermeasures against toxic agents. Prophylactic or preventative countermeasures would take the form of advanced vaccines for common BW agents such as anthrax, botulinum, ricin, Venezuelan equine encephalitis and plague, the development of biological scavengers against nerve agents, and cyanide pretreatment injection. Therapeutic or post-exposure countermeasures include the development of a multichambered auto-injector for battlefield use that would replace the multiple injections against nerve agent contamination that are currently necessary for treatment.

The Nuclear Battlefield

Jimmy Carter, who could never pronounce the word "nuclear" correctly, was nevertheless the president who presided over the largest upgrading of U.S. nuclear forces in history. Prior to the Carter Administration, the doctrine of MAD ruled nuclear policy. MAD stood for "mutually assured destruction." A superpower nuclear exchange would result in a global holocaust and was therefore unthinkable in conventional warfare scenarios: so went the MAD doctrine.

After the Carter-era rethinking of nuclear confrontation, this was no longer the case. Theater nuclear warfare was seen by the United States as a key to deal quickly with the waves of Soviet-built T-series tanks that would herald World War Three. These would roll through the Fulda Gap and inundate Germany and Western Europe unless stopped quickly (although many Europeans saw things differently, claiming the Soviets could be stopped conventionally).

With this change in the SIOP or America's strategic integrated operations plan for nuclear warfare, a gorgon's head of smaller, "cleaner," and lower-yield tactical nuclear devices were built. Unless a truce was called, most wargaming scenarios saw an East-West superpower confrontation in Europe going nuclear in about fourteen days.

With the coming of glasnost and perestroika, the fall of the Berlin Wall, and other such global events, this scenario is now far less likely than it has ever been. Tactical nuclear forces have been rolled back from Western Europe in unprecedented bilateral reductions. On the heels of Russia's 1991 August revolution which defeated a resurgent communist countercoup, massive unilateral rollbacks of ballistic nuclear weapons by the Bush Administration further reduced international Cold War–era tensions.

Nevertheless, nuclear weapons remain in the U.S. arsenal, and plans for their use in certain contingencies do exist, even as new technologies refine their military capabilities. At the same time, the breakup of the former Soviet Union has seen an unprecedented nuclear arms bazaar in which not only weapons-grade plutonium but reportedly warheads have been made available to parties with the hard currency to buy them on the clandestine arms market.

In 1995, German authorities disclosed that an international ring of smugglers had been peddling about nine pounds of Russian plutonium-239, enough to cook a major-league nuke. That same year, the disappearance of stockpiled Soviet theater nuclear missiles, dubbed "Fat Boys" because of their characteristic shape, caused a consternation among intelligence officials. Unlike the plutonium, the disappearance of the Fat Boys is permanent; the missiles have never been seen since.

But it's Russian reactor sites that may prove the ultimate nuclear Sword of Damocles, or Pandora's Box, depending on which fable you prefer. Russia currently has ten major active nuclear reactor sites. Each produces a half-ton of weapons-grade plutonium per year. That's more fissile material than in the combined stockpiles of Great Britain, France and China; enough plutonium for about one hundred one-megaton nuclear warheads. The reactors produce the plutonium as a byproduct of the controlled nuclear chain reactions that generate nuclear energy for electricity, heat and other domestic uses.

Most of this plutonium is stored at the reactor sites, where many workers do not receive pay for months on end. And the fox is watching the henhouse: Russian nuclear installations are guarded by the so-called Special Installation Guards or *Okhrana Osobykh Ob'ektov* (Triple-O's). The ranks of the Triple-O's boast the dregs of the former Soviet military, washouts from Russian combat units with mental or disciplinary problems.

These realities—combined with the grinding poverty of the Russian masses, the widespread corruption of the Russian leadership, and standing offers from Libya, Iran, Iraq and multinational criminal cartels to pay vast sums for supplies of weapons-grade plutonium and technologies for weapons of mass destruction—add up to a nightmare scenario in the making. Tomorrow, we may reap the whirlwind born of the instability caused by dubious Western victory in the Cold War.

Today, a host of small or nonaligned nations have joined the once privileged ''nuclear club'' run by the industrialized Western nations. Israel, Pakistan, India, North Korea, South Africa—to say nothing of China— all are known to have nuclear capability. Iraq is probably on this list as well. It continues to play a cat-and-mouse game with United Nations nuclear inspectors while moving ahead with a secret weapons program to build nukes and other weapons of mass destruction.*

As theater ballistic missile technology continues to proliferate and the accuracy and range of missiles improves, nuclear warheads could be fired at theater targets by a rogue state no longer content with playing Saddam-style rope-a-dope.

This portends that tomorrow's major regional conflict could very quickly go nuclear under certain circumstances. All it might take for such a scenario to materialize would be the launch and detonation of a single nuclear warhead or NBC weapon to trigger a retaliatory strike by either regional or Coalition forces in theater. In this case the fall of one domino could start a chain reaction of appalling consequences for mankind.

A Fast Nuclear Primer

There are two main types of nuclear munitions— fission weapons (A-bombs) and fusion (H-bomb or thermonuclear) weapons. Representative of the first type are the Hiroshima and Nagasaki nukes.

In fission weapons a core of fissionable material— plutonium-239 or uranium-235—just below critical mass (the point where nuclear chain reaction occurs)

---

*In December 1998, Operation Desert Fox conducted hundreds of strikes against military targets in Iraq. These included targets identified as facilities for the manufacture of weapons of mass destruction and facilities for command and control of such weapons. Although there's little doubt it was damaged in the attacks, only wishful thinking would have it that Saddam's banned weapons program was critically hampered.

is imploded to bring it to critical mass. This is accomplished either by the detonation of shaped charges that rapidly compress a lump of the nuclear explosive or by firing lumps at one another at high speed through tubular "guns."

Fusion weapons use a primary fission detonation to trigger the explosive chain reaction. Once this begins, the fusionable material "goes critical" and results in a secondary nuclear explosion. The combined explosive power of the two-stage process can be a hundred times higher than a pure fission weapon alone.

Nuclear groundbursts are useful against hardened missile silos, command centers buried deep underground and similar installations, while airbursts are best used against extended urban areas. In an airburst about one-half of the nuclear energy released by the weapon is transferred to the air in the form of an explosion wave or shock wave that extends in spherical form at supersonic speed from ground zero, the point of detonation.

The shock wave produces a sudden, drastic increase in air pressure that is especially high in the immediate vicinity of the explosion. Simultaneously, there is a brief burst of hurricane-force wind. As the shock wave hits the ground, it creates an air-induced blast effect similar to earthquake. This nuclear one-two punch can cause widespread destruction of large buildings in a matter of seconds.

Nuclear weapons also release a thermal pulse. This pulse creates a blinding flash of light that far surpasses the sun's in brightness and intensity. The nuclear flash is followed by a fireball that radiates intense heat which ignites combustible materials and is able to melt human flesh right off the bones.

Nuclear   detonations   also   produce   nuclear-electromagnetic pulse or N-EMP. This is a tremendous surge of electrical voltage that partially or totally destroys virtually all electronic equipment encompassed within the area of detonation. Within a span of a few nanoseconds, an electrical field of fifty kilovolts or more per minute is created. Airbursts at high altitudes generate especially large-scale EMP effects, which are limited only by the earth's curvature.

Nuclear explosions also produce radioactive fallout. Fallout occurs only when the point of detonation is on the ground. Thousands of tons of soil and rock hurled skyward by the explosion mix with the radioactive fission products of the nuclear reaction during this process.

In twenty-four hours most of the material falls back to earth as a visible rain of dust and ash particles. It's also carried away by the wind, and depending on weather conditions, can be deposited over an area of more than a thousand square miles. Radiation levels decrease rapidly shortly after the explosion, and then more gradually diminish.

Strategic Versus Theater and Tactical Weapons

Nuclear weapons can be classified according to their range and intended targets as either strategic, theater, or tactical weapons.

Strategic nukes are intended to strike targets at large distances and are generally equipped with multimegaton warheads, either MIRVed or unMIRVed. The delivery systems for strategic nukes are either ballistic missiles fired from land or sea or strategic bombers such as the B-52 or B-2 Spirit.

Theater and tactical nuclear weapons have far more limited ranges—from twelve to 125 miles—and are intended for smaller, less distant targets, such as tank formations, military command centers, railway heads, large bridge spans, and dams.

When detonated as airbursts high in the atmosphere, theater and tactical nuclear weapons can also generate devastating N-EMP that can destroy the enemy force's communications capability and even stop the engines of its vehicles and mechanized hardware on the ground.

Such weapons can be delivered by a variety of means, including low-yield nuclear warheads that can be fitted to artillery shells. These can be fired from howitzers to destroy troops and equipment when the enemy is dug-in and a conventional attack is likely to be bloody or has failed to gain a high-priority objective.

The U.S. Department of Defense recently released a report that seemed to indicate a new readiness to use tactical nuclear weapons as bunker-busters. Ever since the Gulf War, our potential adversaries have been burying their C4I installations deeper and deeper underground, where they hope these battlefield command centers will be impervious to the brutal pounding of conventional explosives and surgical strikes by precision guided munitions (PGM) that brought Iraq's military command structure to its knees.

U.S. warplanners seem to have taken notice of this development. Without fanfare, DOD reported that the standard B61 nuclear device had been upgraded for use against protected targets. This revamped nuclear weapon's hardened outer casing penetrates the ground where it detonates with devastating force. Because of its size and weight, the new B61-11 is ideally suited to delivery by the stealthy B-2 Spirit strategic bomber. While treading softly, the Pentagon continues to twirl the big nuclear nightstick.

Special Atomic Demolition Munitions (SADM)

Nukes delivered by air in one form or another will also be complemented by nukes delivered by the grunt on point—tomorrow's soldier will probably carry tomorrow's nuke around with him as his predecessors have carried bazookas, rifles, hand grenades, missile launchers, and canteens. Such next-generation nuclear devices will be smaller and lighter, but no kinder or gentler nukes.

The miniaturization program has produced the SADM (pronounced say-dum). These special atomic demolition munitions are lightweight and small and pack enough nuclear punch to destroy large quantities of unfriendly infrastructure.

There is little in the open literature about SADM programs. In fact, there is virtually nothing. These exotic nuclear development programs are among the most clandestine defense programs today, their operating expenses hidden in line items such as "special updates" in Pentagon appropriations budgets.

From the trickle of information available on the subject, we know that a standard SADM measures about thirty-five inches in width, twenty-six inches in length, and twenty-six in height, contains a fifty-six pound warhead, and packs an explosive wallop of between 0.01 to one kiloton—that is, up to one hundred tons of TNT. Some three hundred SADM weapons are known to be deployed, but the actual numbers are probably much greater than this figure indicates.

For years special forces detachments have been trained to insert clandestinely into enemy territory and emplace these man-portable nuclear devices. At SOF centers at Fort Bragg, North Carolina, SADM training is continuously rehearsed. The SADM capability is considered to be an important mission of special forces, a field manual reveals, "and is constantly exercised in quarterly nuclear surety inspections and field exercise."

"The [SADM] device," the manual continues, "can be deployed either by static line, military free fall, or . . . scuba. Scuba mission training is ongoing within each of the special forces groups."

NATO forces, teams from Britain, Belgium, Greece, Italy, the Netherlands, Turkey, and Germany, have all trained in use of SADM, though it is reasonable to assume that proprietary access to the munitions was strictly controlled by the U.S. Military manuals state that specific targets of SADMs are: "Blocking avenues of approach by cratering defiles and creating rubble, severing routes of communication by destroying tunnels, bridges, canal locks or cratering roads, creating areas of tree blowdown or forest fires, cratering areas subject to hostile airmobile landings; creating water barriers by destroying dams and reservoirs."

By the last years of the Cold War era, U.S. SOF teams were poised to infiltrate East Bloc countries and go to ground once World War Three began. A reconnaissance database built up from years of regular infiltration through sewers and underground tunnels in communist countries contained up-to-date information on which routes inside were open and which were closed.

Special contingency plans called for infiltration, concealment at preplanned hide-sites, and then the planting and detonation of portable nukes to wreck civilian and military targets as soon as a superpower war started climbing the DEFCON ladder—the defense condition threat scale graded from five to one, with five being peacetime and one known as "cocked pistol," the prelude to nuclear war.

Tomorrow's soldier will field nukes much smaller than yesterday's backpack nukes. Unpublished reports substantiated by sources in the defense establishment hold that nuclear weapons capable of delivering close to half a megaton of explosive force have been developed that are nearly the size of a thermos bottle and weigh little more than a laptop computer.

These clean, low-yield "micro-nukes" could reduce a major bridge span to slag in seconds, but would not need to be emplaced in dangerous commando-style operations as is the case with conventional-explosive demolition charges. One small SADM would be the sole capability required to take down a bridge, and the munition would only need to be planted within an approximate square city block of the objective. The SADM could be timed for delayed detonation or be triggered by long-range remote control or various other options.

Space Weapons

The projection of U.S. military power into and the deployment of weaponry from space has sometimes been trumpeted and other times whispered by warplanners, but it has been an obsession since the close of World War Two, when Adolf Hitler's primitive rocketry programs prefigured the devastating military potential of space weapons.

The existence of the Nazi V-2 "Vengeance Weapon" in the final months of the war shocked the Allies, who, aided by the unsung exploits of the hundreds of slave laborers who deliberately miswired the missiles' guidance systems, unleashed massive bombing raids to destroy launch and construction facilities at Nazi bases in the German Baltic.

After the war the scientists responsible for Nazi rocketry formed the core around which the space programs of the U.S. and Russia coalesced, and it's a historical sidelight of some importance that the V-2 rocket technology these men pioneered is essentially the same as the Vengeance Weapon's modern-day counterpart, the Scud.

Throughout the Cold War, the United States fielded various initiatives to place military assets into orbital space. These included the X-15 and Dyna-Soar rocketplanes of the 1950s and 1960s and the Space Shuttle first launched in the 1970s, part of whose mission was to loft military satellites into orbit, return for repair and refueling of defense assets in place, and test and deploy new orbital systems.

Since the late 1960s the emphasis on space-based military assets has been on passive systems—that is, those used for surveillance and communications activities. But for reasons cited earlier in this chapter, this state of affairs is bound to change as the next century unfolds.

There are two missions that will drive the development of space-based weapons systems. The mission is first to protect the satellite communications and surveillance assets without whose presence both peace and war in the New World Order could not be sustained, and second, to protect against theater nuclear attack, undertaking some of the duties of the old SDI shield concept. This mission is a defensive one.

The second mission is offensive in nature. It will see the development of weapons intended to strike preemptively at targets on earth and in space.

Critical to both these missions is another throwback to the early days of space exploration and militarization, something not heard much about since the early 1960s—manned military orbital platforms, or space stations, as they have come to be popularly called. These would be used for both defensive and offensive activities, both carried out from the station itself or from arrays of SDI-type weapons under the control of the station, primarily laser, kinetic, and conventional explosive in nature.

For offensive purposes, an excellent weapon to be deployed from an orbital space station would be a non-nuclear EMP beam generator, of a type already under development at applied physics laboratories serving military research and development. N-EMP, or nuclear-electromagnetic pulse, is a wave of high-energy electrons that is produced by nuclear explosions. It has the effect of killing electrical and electronic systems unless these are specially protected from N-EMP effects.

One of the strategies for destroying satellites in orbit used by Russian antisatellite satellites is to detonate a low-yield nuclear explosive. This produces N-EMP capable of destroying all onboard systems, but because it detonates in orbit, there is no earthbound radiation. A larger nuclear detonation or one in a lower orbit could extend N-EMP effects to airborne or earthbound electrical or electronic systems. In fact, it is said that a pet idea of General Schwarzkopf was to knock out all Iraqi C4I in one fell swoop with just such a tactic during Desert Storm.

A better way to do this is to generate tactical non-nuclear EMP using beam generators with one-gigawatt or higher outputs. The problem with this approach is that with current or even near-future technology there is simply no way to generate the several megajoules of energy that are required in the few milliseconds necessary to create an EMP pulse.

A large bank of storage capacitors must first be charged by the beam generator and then release the charge in a single burst. Even with advances in superconductors which would increase the amount of energy generated and reduce storage capacitor size, a tactical EMP system would still have to be fairly large—say, the size of a small truck.

Such a system would be cumbersome on earth, but would be much more maneuverable in space, where it could also be easily assembled from parts lofted into orbit by the Space Shuttle or another reusable manned orbiter. Located on a manned military space station, a battery of EMP guns could be maneuvered into position above the scene of a major regional conflict, precision-targeted, and then fired.

A single burst of beam-generated EMP would knock out any communications, computers, navigation, data processing, and even electronic automotive ignition systems that were not specially protected below, and precision-guided munitions could deal with those that the EMP barrage did not destroy. As a soft-kill option, tomorrow's space station equipped with EMP capability could conceivably stop tomorrow's Saddams in their tracks very early on in a military confrontation.

Information Warfare (IW):
The Microchip as Aimpoint

Information warfare (IW) has become a cornerstone of military planning, almost an obsession, possibly because it has proved as difficult to define as the stealth aircraft that dropped their bombs on Baghdad were difficult to see by Iraqi gunners. In both cases, blind shots in the dark have been taken in attempts to nail down the bogie which, also in both cases, has proved itself a most elusive target.

One good definition of IW is the USAF's working definition as "any action to deny, exploit, corrupt, or destroy the enemy's information and its function as well as protecting against those actions and exploiting friendly information operations."

Another excellent definition of IW, really more of a metaphor, was one proposed at the TechNet '95 conference in Washington by USAF Major General Ken Minihan, Assistant Chief of Staff for Air Force Intelligence. Minihan suggested that IW be thought of as "the microchip as aimpoint," adding, "I want to defend it and I want to attack it, and I want to do that in an integrated way."

Information warfare, then, treats the microchip as aimpoint. As with any other military or civilian target, it can be be attacked and it can be defended from attack, and attack and defense can take either high-tech or low-tech forms. Sophisticated electronic jamming and the explosion of a dumb iron bomb to neutralize a C4I node, for example, would both be forms of information attack by this definition.

In this way the microchip is the same as virtually any target. But in a number of other ways it's very different in that it's not the chip per se that is important so much as the information it contains and processes.

As we've seen throughout this book, information has become critical to the performance of virtually every combat system that will fight tomorrow's war. This makes the microchip the Achilles' heel of all of those systems. In the words of the USN's Deputy Director of Intelligence, Admiral William Studeman, "Massive networking makes the United States the world's most vulnerable target for information warfare."

Destroy an air combat wing and other warplanes can be put in the air, but destroy a critical information linkage and nothing flies. Take a regiment prisoner and fresh troops can be brought in, but disrupt the flow of tactical information and those same troops can be rendered as impotent on the battlefield as blind men would be. Or take the following scenario.

Hackers have penetrated the computer databases containing design information for a new generation of advanced tactical fighter being developed by a major aerospace corporation. The hackers leave behind a sophisticated Trojan Horse (TH), a viral program (or CMC, for "computer malicious code") that goes unrecognized as work proceeds on the plane's flight control system.

When the integrated circuit chips that comprise the heart of that system are ultimately manufactured, they contain the CMC. It lies dormant until the planes are flown in combat against the military adversary for whom the hackers had worked. Then, in response to specially coded transmissions by enemy forces, the Trojan Horse's "trap door" opens to release the digital info-weapons hidden inside, lines of destructive viral code.

The first thing this rogue code does is render the ultrastealthy warplanes visible to enemy radars. The second thing it does is enable the enemy to make the planes' flight control systems fail. One by one, the planes lose control, plunge to earth, and explode into fireballs, destroying U.S. air superiority and compelling allied forces to wage a ground war on the enemy's terms in which thousands of troops die.

This scenario is fiction, because it hasn't happened and may never happen. But another example of an enemy nation waging what might rightly be regarded as information attack on a strategic target in the United States did happen. In 1993, Iranian-backed terrorists detonated a crude but effective bomb at the World Trade Center (WTC) in New York City that caused severe damage. Militarily, this attack could be viewed as attack on a strategic information target, since the WTC housed major telecommunications switching nodes and computer systems affecting Wall Street trading, and thus affecting the global marketplace.

In fact, it has been estimated that of the some one hundred nations said to possess an information attack capability against the United States, some fifty of these nations have either already done so or are in the process of doing so, mostly in the form of unauthorized break-ins of government and commercial computer networks by hackers in the pay of foreign governments. One such break-in involved the penetration of Rome Laboratory, a civilian research and development facility for largely classified weapons systems administered by the USAF in 1994 in which secret files were downloaded from the facility's computers.

Drawing a parallel to the revolution in military af-

fairs (RMA) we've already discussed, some observers
have called the risks that information-based systems
pose to our ability to successfully wage war in the
coming years as a "revolution in security affairs"
(RSA), likening this destabilizing situation to that prevailing at the start of the nuclear arms race and calling
for the implementation of an information warfare single integrated operations plan (SIOP). Some have
gone so far as to warn of an "electronic Pearl
Harbor" scenario, where a preemptive info-attack on
U.S. strategic and tactical computer systems destroys
America's capability to go to war against its next adversary before the first troop ships have even left port.

Just as likely, if not more so, is the future exploitation of IW capability by the U.S. to achieve information dominance (ID) in the digital battlefield of
tomorrow's war by suppressing the information resources of opposition forces and protecting those of
friendly forces. In some cases, information warfare
might even be used as a replacement for the regional
presence of physical military forces.

A DOD task force on information warfare has
stated that "on occasion, information alone may be
enough to attain U.S. objectives" in combat by exploiting the potential of virtual or synthetic battlespace, a concept that should be familiar from an
earlier chapter. Since the enemy sees, hears, and perceives the warfare environment through the network
of computerized infosystems that form the nervous
system of the digitized battlefield, a deception capability acting on that system could cause severe dis-
ruption of the enemy's ability to fight. This form of
IW attack could take many forms, from creating nonexistent wings of fighters on enemy scopes to planting

CMCs on his computers that would cause his tanks to see ours as friendlies.

Alternatively, forms of IW could be deployed directly against the soldier-in-the-loop—that is, against the human troops on the ground instead of the machinery that tomorrow's enemy soldier controls and with which he interacts. Visual stimulus and illusion (VSI) is a form of virtual reality warfare (VRW), which is in turn a subset of information warfare.

One form of VSI might employ a technology known as the Bucha Effect, which uses arrays of high-intensity strobes and other light sources that flash at specific human brainwave frequencies to distort brain activity, producing extremely disorienting hallucinations. The Bucha Effect, incidentally, has been pioneered by the Russians, who are said to have experimentally refined Bucha technology to a high degree of sophistication.

Another form of VSI warfare would use multimedia laser holograms to project large-scale, three-dimensional depictions on the battlefield of anything from an onrushing column of nonexistent hypertanks to mythical gods in the heavens. Picture how effective this type of VSI attack on Iraqi troops in the Gulf War might have been if only a single wave of high explosive bombing were followed up with a gigantic hologram of both Saddam Hussein and Allah himself instructing the enemy to lay down his weapons and assume the "dying cockroach" position.

The list of future information warfare applications goes on, as does the attempt to come to grips with both the dangers and opportunities for the future presented by the new reliance on military information systems, and to define a comprehensive definition of IW as well as a national policy in support of it.

Two years-long studies by the U.S. government have recently published their findings in the form of two reports. Critical Foundations was a study conducted by the President's Commission on Critical Infrastructure Protection, issuing its report in October 1997. Transforming Defense, a report of the National Defense Panel, was issued in December 1997. Both studies have cited the pressing need to address the critical role of information in attack and defense in the first decades of the next century.

If the microchip is aimpoint, then all of us stand squarely behind it.

We Are All the Target:
The End of the Home Front

We have now returned full circle to the premise stated at the start of this book.

Tomorrow's soldier will be all of us because the home front has been rendered obsolete by military technology and political agendas. Tomorrow's battlefield will extend from the distant combat zone of the MRC to the towns and cities of America itself. Yesterday's complacency regarding foreign attack will prove a dangerous illusion in the coming years.

Throughout the 1980s, the continental United States considered itself immune to European-style terrorism that had resulted in extensive damage to property and life in foreign settings. When, during the Gulf War, Saddam Hussein threatened to activate terrorist cells in the United States to wage a behind-the-lines assault on civilian and political targets, the FBI and other counterterrorist agencies in the United States took the threat seriously and issued domestic advisories and warnings, but few ordinary citizens took these seriously. In major American cities, which would have been the first targets of attack, business went on as usual as the Gulf War riveted millions to their television screens.

Saddam's terrorist cadres never materialized and things went back to normal after the war—the complacent "it can't happen here" attitude seemed again to have been justified. But a few years later, the World Trade Center (WTC) bombing shattered this bubble of indifference forever, engulfing a symbol of American business power in flames and choking smoke, and almost bringing New York City to a standstill.

The vulnerability of everyone to terrorist attack was brought home to this author, who after the bombing, realized that he had come close to being a casualty of the WTC blast. About a week before the explosion I had parked a rented car in the underground garage beneath the Hotel Vista, probably not far from where the van with home-brewed explosives that caused the damage had been parked.

As I drove down the access ramp, then into the rental car parking garage directly beneath the lobby of the hotel, which was damaged by the blast, I saw no one in sight. I parked the car in the silent, empty garage, then went upstairs to the lobby, still meeting no one. At the desk of the car rental company I handed the keys to a smiling company rep. Had my trip been timed a mere week later, I could literally have been blown to bits when the explosive-packed van exploded.

But this is not the only coincidence. I later learned that the terrorist group alleged to have planted the bomb owned a large compound within a ten-minute drive from the remote farmhouse in upstate New York

I had driven my rental car hundreds of miles to look at in an effort to "get away from it all." Fortunately, I hadn't been overwhelmed by the property.

More recently, residents of New York City learned how close they had come to a possibly even more horrifying "in-your-face" form of terrorist attack. In August 1997, a Palestinian immigrant, Abdul Mossabah, admitted to police that he was part of a terrorist cell that was preparing to suicide-bomb the New York City subway system.

The group had been building pipe bombs from off-the-shelf components and stockpiling the weapons in readiness for the planned suicide bomb attack. The strike was to take place the following day, during the morning rush hour; the members of the cell had already prepared their suicide notes, laced with defiant words against America.

Plans called for the suicide squad of human bombs to board separate trains and at a suitable point open their coats and shout "Allah Akbar!" then blow themselves and the subway passengers around them to smithereens.

But at this writing, only a matter of months after the plan hit national headlines, an attack potentially far more disruptive than the WTC bombing has already been forgotten. The "it can't happen here" complacency has already set in. But it is certain that other terrorist cells are at this moment planning other assaults, and the chilling likelihood is that some of these will succeed later, if not sooner.

During the next MRC, it would be wise to take precautions, whether or not the spokesman for the enemy announces his regime's intention to strike at the U.S. home front as well as at American troops in the combat theater. In tomorrow's war, it is likely that the threat of retaliatory terrorist attacks on America will be no idle boast.

Because tomorrow the enemy will have better, more sophisticated weapons available to it, and the attack will probably be a high-visibility strike on a large scale of devastation.

It's possible that some form of modified NBC weapon could be used, such as a "dirty" subkiloton nuclear device that spreads massive amounts of radioactivity. Or, again, chemical or biological agents could prove to be the weapons of choice. These would make ideal weapons to be leveled against large cities because of the high kill probability in such a dense target population.

Air-dispersed neurotoxic agents that work quickly could cause massive damage to life and property. Another possibility would be for a terrorist equipped with a small nuclear device to detonate it in an aircraft passing over the center of Manhattan, Chicago, or Los Angeles.

Such a nuclear airburst would release enough N-EMP to disrupt severely the flow of transit, commerce, and municipal services for miles around and create a state of severe civil disorder. Bank accounts would be wiped out in the blink of an eye. Cars, trucks, and passenger buses would crash and cause massive gridlock, their electronic ignition and control systems knocked out by the high voltage pulse. Planes would tumble from the sky or careen into skyscrapers as navigation systems went blind and stopped functioning. If combined with a chemical attack, such a terrorist strike could kill millions of noncombatants in a matter of minutes and cripple the target cities for years to come.

Rogue states with the incentive, the will, and the

means to attack the continental United States need not only depend on agents-in-place to plant and detonate explosive devices of all types. In an age of computers married to high-energy munitions, both conventional and nuclear, he can strike from outside the sanctuary of our national borders, too.

The downing of TWA flight 800 over Long Island Sound only some twelve minutes after it had departed from New York's JFK International Airport en route to Paris in the summer of 1996 illustrates how vulnerable we all are to a terrorist strike from coastal waters.

Whatever the official explanations of the aircraft downing, the question of whether the crash occurred as a result of a missile strike can never be fully discounted. Using a stolen man-portable rocket launcher firing an antiaircraft round such as Stinger, terrorists onboard a vessel in the waters of the Sound could have downed the jet which was flying at low enough altitudes and speeds during that part of its ascent to be vulnerable to attack.

Similarly, as cruise missile technology becomes increasingly more available to rogue states, and the range and killing power of such weapons increases, the threat of a rocket strike from a ship lying off coastal waters will dramatically increase.

Other, more technologically advanced states, such as China and Korea, are already close to fielding mobile ICBMs whose range could threaten America's West Coast; General Xiong Guangkai, a senior Chinese military officer, has stated that China already possesses the capability of reaching Los Angeles with a ballistic missile strike. And while engaged in an effort to make Libya a tourist oasis by the year 2000, its despotic leader Mu'ammar al-Gadhafi also talks and dreams of acquiring strategic nuclear missiles capable of striking New York City.

Armed with a chemical, biological, or nuclear warhead—an advanced ICBM could carry a MIRVed payload laden with grapefruit-sized submunitions packed with chemical or biological warfare agents—such a missile could do tremendous damage, and it would require only a single warhead to devastate completely a major city such as New York or Los Angeles.

With tomorrow's war, the home front will no longer exist in its past form. There will no longer be safe sanctuaries anywhere. The risks of the battlefield will be shared by one and all, because tomorrow's soldier will be everyone, everywhere.

Frontline troops will be better trained, better protected, and better armed. Advances in digital information technology coupled with stealth and speed, with mobility and precision, and with jointness and interoperability, will increase the range and striking power of combat forces to unprecedented levels. But unfriendly forces will also share in this revolution in military affairs, and they, too, will benefit from the proliferation of new combat technologies.

If war is politics by other means, then our surest strategy for a winnable war is never to have to fight one again. Because this may prove as impossible in the future as it has proved in the past, the next best option open to America and her allies is to continue to invest in the development of new military technologies, force structures, operational concepts, and the geostrategic policies necessary to support these innovations.

The payoff of this investment may be a dividend

of continued global peace and comparative international stability in which tomorrow's soldier will play
a critical part, wherever he may serve, in protecting
against those bent on having it otherwise.

Appendix: Acronyms

AAM air-to-air missile
AAR air-to-air refueling
ABL airborne laser
ACDS advanced combat direction system
ACTD advanced concept technology demonstration
AFMSS Air Force mission support system
AFS advanced fire support
AMRAAM advanced medium-range air-to-air missile
APFSDS armor piercing fin-stabilized discarding sabot
AR augmented reality
ASAT antisatellite
ASEAN Association of Southeast Asian Nations
ATA/D aided target acquisition/designation
ATGM air-to-ground munition/missile
ATT antitraction technology
AWACS airborne warning and control system
BAT ballistic antitank
BLAM barrel-launched adaptive munition
BLISK bladed disk
BM/C3I battle management+ C3I
BUR Bottom-Up Review
BW biological warfare
C2 command and control

C3 command, control, and communications

C3I command, control, communications, and intelligence

C4I command, control, communication, computers, and intelligence

C4IFTW C4I for the warrior

C4ISR command, control, communications, computers, intelligence, surveillance, and reconaissance

CAP combat air patrol

CAT combustion alteration technology

CAWS close assault weapon system

CBIRF Chemical Biological Incident Response Force

CCD charge coupled device

CIC combat information center

CIS Commonwealth of Independent States

CIWS close-in weapon system

CMC computer malicious code

CODA chemical/biological operational decision aid

COIL chemical oxygen iodine laser

COTS commerical off-the-shelf

CRT cathode ray tube

CWC Chemical Weapons Convention

DARPA Defense Advanced Research Projects Agency

DEFCON defense condition

DIA Defense Intelligence Agency

DIRCM Directed Infrared Countermeasures system

DOD U.S. Department of Defense

DP Distant Portrait Program

DSWA Defense Special Weapons Agency

DUF deep underground facility

DVBX digital video branch exchange

ECAD Equipment du Combattant Debarque

ECCM electronic counter-countermeasures

ECM electronic countermeasures

EMP electromagnetic pulse
EOB electronic order of battle
ERA explosive reactive armor
ERP evoked response potential
EW electronic warfare
FAAD forward antiaircraft defenses
FARV future armored resupply vehicle
FBE fleet battle experiment
FFG force feedback glove
FIFV future infantry fighting vehicle
FLIR forward looking infrared
FSU Former Soviet Union
GCCS global command and control system
GIGN Groupement d'Intervention de la Gendarmerie Nationale
GLO-MO global-mobile
GPS global positioning system
GSM ground station module
HAE UAV high-altitude endurance unmanned aerial vehicle
HAHO high-altitude, high-opening
HARM high-speed antiradiation missile
HDBTDC hard and deep buried target defeat capability
HEAT high explosive antitank
HMD helmet mounted display
HMMWV High Mobility Military Warfare Vehicle
HOTAS hands-on throttle and stick
HUD head-up display
ICBM intercontinental ballistic missile
ICNIAS integrated communications, navigation, and identification avionics system
ID information dominance
IFF identification friend or foe
IHAVS integral helmet audio-visual system

ILS integrated large screen display
INS inertial navigation system
IR infrared
IRIN Islamic Republic of Iran Navy
IT information technology
IW information warfare
JDAM joint direct attack munition
JFC joint force commander
JIRD joint initial requirements document
JMCIS joint maritime combat information system
JSLIST Joint Service Lightweight Integrated Suit
    Technology
JSF joint strike fighter
JSOW joint standoff weapon
JSTARS joint strategic tactical airborne radar system
KE kinetic energy
KEAS kinetic energy antisatellite system
LCAC landing craft assault craft
LEL low energy laser
LHN long haul network
LHX light helicopter, experimental
LIDAR light detection and ranging
LMRS long-term mine reconnaissance system
LO low observable
LOS line of sight
MAD mutually assured destruction
MALD miniature air-launched decoy
MARS Mine Actuation Registration System
MAV micro air vehicle
MBT main battle tank
MCC management command and control
METT-T mission, enemy terrain, troops available, and
    time
MICV mechanized infantry combat vehicle
MiG Mikoyan-Gurevich

MIRV multiple independently targetable reentry vehicles

MLRS multiple-launch rocket system

MM millimeter

MOBA military operations in built-up areas

MOD British Ministry of Defense

MPP massively parallel processor

MRBM medium-range ballistic missile

MRC major regional conflict

MRF-MiG multirole fighter

N-EMP nuclear-electromagnetic pulse

NATO North Atlantic Treaty Organization

NBC nuclear biological chemical

NBC/M NBC weapons and the means to deliver them

NDT nonlethal disabling technologies

NMRS near-term mine reconnaissance system

NRT near real-time

NUWC Naval Underwater Warfare Center

NVPS night vision piloting system

OCU operator control unit

OODA observe, orient, detect, act

OOTW operations other than war

OPFOR opposition forces

OPS operations per second

ORD operation requirement document

PDW personal defense weapon

PFU Patriot fire unit

PGM precision-guided munition

PRC People's Republic of China

QDR Quadrennial Defense Review

R&D research and development

RAM radar-absorbant materials

RCS radar cross-section

RMA revolution in military affairs

RNIP-Plus Ring Laser Gyro/Global Positioning System Navigation Improvement Program

ROV remote-operated vehicle

RSA revolution in security affairs

RSO radar system officer

RST reconnaissance, surveillance, and targeting

RSTA reconnaissance, surveillance, and target acquisition

SADARM sense and destroy armor

SADM special atomic demolition munition

SAM surface-to-air missile

SAR synthetic aperture radar

SAS Special Air Service

SDF Japanese Self Defense Forces

SDI Strategic Defense Initiative

SDV swimmer delivery vehicles

SEAD suppression of enemy air defenses

SGR sortie generation rate

SINCGARS single-channel ground and airborne radio system

SIOP single integrated operations plan

6DOF six-degree-of-freedom

SOF special operations forces

SOFPARS special operations forces planning and rehearsal system

SQUID superconducting quantum interference device

SRBM short range ballistic missile

SSBN nuclear ballistic missile submarine

STOL short takeoff and landing

STOVL short takeoff/vertical landing

STOW synthetic theaters of war

STRATCOM Strategic Command

SU Sukhoi

TACAIR tactical air forces

TBA tactical battle assistant

TBM tactical ballistic missile or theater ballistic missile

TH Trojan Horse

THAAD theater high-altitude area defense

THEL tactical high-energy laser system

TLAM Tomahawk land attack missile

TO teleoperation

TOW tube-launched, optically tracked, wire-guided missile

TP telepresence

TR televisual reality

TRADOC training and doctrine command

TUV tactical unmanned vehicle

UAV unmanned aerial (or aerospace) vehicle

USAF United States Air Force

USMC United States Marine Corps

USN United States Navy

USSOCOM U.S. Special Operations Command

UUV unmanned underwater vehicles

V-2 Vengeance Weapon

VCASS visually coupled airborne systems simulator

VHSIC very high speed integrated circuit

VRW virtual reality warfare

VSI visual stimulus and illusion

WSO weapon system officer

WTC World Trade Center

Selected Glossary of Terms

Advanced tactical fighters (ATF) refers to a new generation of fighter planes that is currently being built and fielded by several nations, including the United States. All these planes share certain common design features, especially the requirement that they can fly at supersonic speeds in "dry" mode, that is, without going to afterburner. Advanced avionics (aviation electronics) or control interfaces between pilot and plane, such as head-mounted displays (HMD), are also integral to the ATF concept.

Air dominance fighters are fighter planes such as the F-22 that are designed to sweep the skies completely of enemy aircraft. Air superiority fighters, such as the F-16, are intended to suppress enemy aircraft activity but also perform a number of other roles, including tactical bombing.

The Aurora Project is what is believed to be a classified program to develop high-altitude transonic stealth aircraft. It is believed that the project has already built manned and unmanned space-capable planes able to fly at better than five times the speed of sound, and that these are already being used to perform high-altitude surveillance missions.

BLAM is the barrel-launched adaptive munition, which like the above program, could be either an antisatellite or antimissile system, capable of killing Scuds and orbital spy platforms alike.

C4I for the warrior (C4IFTW) is a set of technologies intended to afford the dismounted infantry soldier with a command, control, communications, computing, and intelligence (C4I) "cocoon" that will enhance his ability to understand the combat environment, aim and fire his weapons, and communicate with both forces in the immediate zone of operations and distant command centers behind the front lines. C4IFTW technologies may combine to produce the "supertrooper" of twenty-first-century land warfare.

The digital battlefield is a concept relying on advanced digital information networks to rapidly bring "sensor-to-shooter" data to military forces as the shape of the battle develops. It is related to the concept of the combined arms battlefield, in which joint interaction of air, sea, and ground forces enables a multilayered, multidimensional approach to waging war.

The Eurofighter is another advanced tactical fighter development program that has been ongoing for several years. The Eurofighter is being developed by a partnership including the United Kingdom, Germany, Italy, and Spain and is intended to produce air force and naval variants of the plane for service in all the participating countries.

Explosive reactive armor (ERA) is applied in applique plates to vulnerable areas, such as the front glacis of main battle tanks.

The "Fritz" helmet is an exact replica of the German *Wehrmacht* helmet of World War II, albeit in Kevlar instead of steel.

IRIN stands for *Niruye darya-iye jumhuriye Islami Iran (Nedaja)*, known to the West as the Islamic Republic of Iran Navy. IRIN has a personnel roster of between 38,000 and 40,000 men, some 18,000 of which are professional seamen.

Low observable (LO) technology, otherwise known as "stealth," is a combination of methods designed to defeat radar identification of combat aircraft by reducing that aircraft's RCS or radar cross section.

Mirage 2000 is a warplane that will be likely to fly in tomorrow's air combat theater, in the hands of friendlies and adversaries alike. The Mirage 2000 is a multirole fighter capable of being fitted with sensitive radars and of carrying a wide variety of missiles, including AMRAAM.

N-EMP, or nuclear electromagnetic pulse, is a wave of high energy that effects electrical and electronic systems unless these are specially protected or hardened against the effects. Non-nuclear EMP can also be produced for offensive purposes using particle beam generators.

Order of battle refers to the disposition of all military forces in the theater of war and defines all practical aspects of how these forces will be deployed and in what relation to one another during the course of their participation in the battle. The electronic order of battle (EOB) is the above definition translated to electronic warfare components of the battlespace.

Revolution in Military Affairs (RMA) is a kind of military perestroika, a new thinking that began to take shape after the breakup of the Soviet Union changed the bipolar superpower relationships of the Cold War into the multipolar regional relationships of the New World Order. This thinking solidified into doctrine in the wake of the War in the Gulf. Prior to these revolutionary developments a completely different set of suppositions about future war prevailed.

Soft Kill is a strategy based on a doctrine whose object is not to annihilate the enemy but to knock him out and then make peace. Its objective is to destroy the enemy's ability to wage war by annihilating its warfighting infrastructure: to defang the serpent and render it harmless.

The TOW missile is a wire-guided missile that needs a direct line-of-sight link between shooter and target for at least three seconds. While the TOW round is in flight, the shooter must keep it fixed in his target reticle while signals transmitted along the wire paying out from its rear keep it on course.

Vetronics is to vehicles what avionics is to aircraft. Like first-line military aircraft systems, tanks, armored personnel carriers and other combat vehicles are often equipped with digital display screens, HMDs and other similar control interfaces.

Virtual reality (VR) is a technology that was developed by the military as a command and control interface between the soldier-in-the-loop and increasingly sophisticated weapons systems.

Weapons of mass destruction (WMD) are weapons capable of maximum destruction of human lives and physical infrastructure over a wide area. Nuclear munitions, high-energy conventional munitions, and chemical-biological warfare agents are all examples of WMDs.

Recommended Reading and Research Sources

Books

WARS: PAST PRESENT AND FUTURE

Crusade in Europe. Dwight D. Eisenhower; Doubleday & Company, 1948

Crusade: The Untold Story of the Persian Gulf War. Rick Atkinson; Houghton Mifflin, 1993

Desert Storm. Time Warner Publishing, 1991

Future War. Frank Barnaby, (Ed.), Facts on File, 1984

The Future of War. George and Meredith Friedman; Crown, 1996

The Mask of Command. John Keegan; Viking, 1987

The Outlaw State. Elaine Sciolino; Wiley, 1991

Second Front. John R. Macarthur; Hill and Wang, 1992

The Second World War. Winston S. Churchill; Houghton Mifflin, 1951

The United States Navy in World War II. S. E. Smith, (Ed.); Ballantine, 1978

War and Anti-War. Alvin and Heidi Toffler; Little, Brown, 1993

War Zones. Jon Lee Anderson and Scott Anderson; Dodd, Mead & Company, 1988

The World Atlas of Revolutions. Andrew Wheatcroft; Simon and Schuster, 1983

STRATEGY AND TACTICS

Brute Force: Allied Strategy and Tactics in the Second World War. John Ellis; Andre Deutsch Limited, 1990

Fighter Combat. Robert L. Shaw; Naval Institute Press, 1985

Fleet to Fleet Encounters. Eric Grove; Arms and Armour Press, 1991

On War. Karl Von Clausewitz, Michael Howard and Peter Paret, (Trans.); Princeton University Press, 1976

Sun-Tzu: The Art of Warfare. Roger Ames (Trans.); Ballantine Books, 1993

WEAPONS AND SOLDIERING

The Arms Bazaar: From Lebanon to Lockheed. Anthony Sampson; Bantam, 1978

Aurora: The Pentagon's Secret Hypersonic Spyplane. Bill Sweetman; Motorbooks, 1993

The Commandos. Douglas C. Waller; Simon & Schuster, 1994

The Complete Book of Fighters. William Green, Gordon Swanborough; Barnes & Noble Books, 1998

Elite Fighting Units. David Eschel; Arco Publishing, Inc., 1994

A Higher Form of Killing. Robert Harris and Jeremy Paxman; Hill and Wang, 1982

It Doesn't Take a Hero. H. Norman Schwarzkopf; Bantam, 1992

Military Small Arms of the 20th Century. Ian Hogg and John Weeks; DBI Books, Inc., 1998

My American Journey. Colin Powell; Random House, 1995

The Naval Institute Guide to the Ships and Aircraft of the U.S. Fleet. Norman Polmar; Naval Institute Press, 1998

Russia's Top Guns. Gallery Books, 1990

Smart Weapons. McDaid and Oliver; Barnes & Noble Books, 1997

SR-71 Blackbird in Action. Squadron/Signal Publications, 1982

Storm Command: A Personal Account of the Gulf War. General Sir Peter de la Billière; Harper-Collins, 1992

When the Pentagon Was for Sale. Andy Pasztor; Scribner, 1995

GENERAL REFERENCE

Assignment Pentagon. Major General Perry M. Smith, USAF (Ret.); Pergamon-Brassey's International Defense Publishers, 1989

International Defense Equipment Catalogue. Munch, 1998

Modern Military Dictionary. Maher S. Kayyali; Hippocrene Books, 1991

The Language of Nuclear War. Semler, Benjamin and Gross; Perennial Library, 1987

Top Secret: A Clandestine Operator's Glossary of Terms. Bob Burton; Berkley, 1986

War Slang. Paul Dickson; Pocket Books, 1994

Periodicals

Air Power International
Army Times
Aviation Week
Defense News
Military History
Military Technology
Modern Simulation and Training
Proceedings
Special Forces Magazine

Websites

GOVERNMENT

Central Intelligence Agency
http://www.odci.gov/cia

Israeli Foreign Ministry
http://www.israel-mfa.gov.il

Japan Defense Agency
http://www.jda.go.jp

Ministry of Defence, United Kingdom
http://www.mod.uk

National Security Agency
http://www.fas.org/irp/nsa

Russia on the Net
http://www.ru

MILITARY

Air Force Association
http://www.afa.org

Defense Advanced Research Projects Agency
   (DARPA)
   http://www.arpa.mil

Joint Chiefs of Staff
http://www.dtic.mil/jcs

NATO
http://www.nato.int

Naval Undersea Warfare Center
http://www.npt.nuwc.navy.mil

STRICOM (U.S. Army)
http://www.stricom.army.mil

The Israel Defense Forces (I.D.F.)
http://www.idf.il

U.S. Air Force
http://www.acc.af.mil

U.S. Navy
http://www.ncts.navy.mil

COMMERCIAL

Aerospatiale
http://www.aerospatiale.fr

British Aerospace
http://www.britishaerospace.com

Defense News Online
http://www.defensenews.com

Lockheed Martin
http://www.lmco.com

Index